Memoirs of the American Mathematical Society

Number 256

David A. Stone

The exponential map at an isolated singular point

Published by the

AMERICAN MATHEMATICAL SOCIETY

Providence, Rhode Island, USA

January 1982 · Volume 35 · Number 256 (second of 3 numbers)

Subscriptions and orders for publications of the American Mathematical Society should be addressed to American Mathematical Society, P. O. Box 1571, Annex Station, Providence, R. I. 02901. *All orders must be accompanied by payment.* Other correspondence should be addressed to P. O. Box 6248, Providence, R. I. 02940

MEMOIRS of the American Mathematical Society (ISSN 0065-9266) is published bimonthly (each volume consisting usually of more than one number) by the American Mathematical Society at 201 Charles Street, Providence, Rhode Island 02904. Second Class postage paid at Providence, Rhode Island 02940. Postmaster: Send address changes to Memoirs of the American Mathematical Society, American Mathematical Society, P. O. Box 6248, Providence, RI 02940.

CONTENTS

Library of Congress Cataloging in Publication Data

Stone, David A.
The exponential map at an isolated singular point.

(Memoirs of the American Mathematical Society, ISSN 0065-9266 ; no. 256)
Bibliography: p.
1. Geometry, Riemannian. 2. Singularities (Mathematics) 3. Mappings (Mathematics)
I. Title. II. Series.
QA3.A57 no. 256 [QA649] 510s 81-19100
ISBN 0-8218-2256-X [516.3'73] AACR2

Research supported in part by N.S.F. grant No. MCS-7802147 and in part by a grant from the City University of New York PSC-BHE Research Award Program.

Abstract

This work is a study in the local differential geometry of a metric space M which has an "isolated conical singularity" at a point P; this means that $M^o = M - \{P\}$ is a Riemannian n-manifold, and that near P the metric is, in a rather technical sense, approximately that of a cone from P on a closed, compact (n-1)-manifold. The main results are that M has a "tangent metric space" T_PM at P which is a cone, and that there exists an "exponential map" at P: a homeomorphism $\exp_P\colon T_PM \to M$ defined near P such that for every $\hat{X} \in T_PM$ near P, the path $\gamma(t) = \exp_P(t\hat{X})$ is defined for $t \in [0,1]$, $\gamma(0) = P$, $\gamma(0,1]$ is a geodesic in M^o, and the "tangent direction" of γ at P is $\hat{X}$.

The most important examples are complex analytic hypersurfaces. Let $M^n \subseteq \mathbf{C}^{n+1}$ have equation $f(x_1,\ldots,x_{n+1}) = 0$, and assume that the origin P is in M. When n = 1, P is an isolated conical singularity of M, so $\exp_P$ exists. When n > 1, let g be the initial homogeneous polynomial of f. P is called "unbranched" if the vector field $\left(\frac{\partial g}{x_1},\ldots,\frac{\partial g}{x_{n+1}}\right)$ has an isolated zero at P. In this case T_PM is homeomorphic to M near P and every geodesic γ from P has a tangent direction $\hat{\gamma} \in T_PM$. If in addition the sectional curvature is bounded above near P in M - {P}, then P is an isolated conical singularity, so $\exp_P$ exists. When n > 1 there are examples of an isolated singularity P which fails to be unbranched, such that there can be no exponential map at P.

The definition of an isolated conical singularity starts with an approximation to the desired exponential map, namely a homeomorphism $h\colon cL \to M$ of the cone on an (n-1) manifold into M such that h(c) = P. Several conditions are imposed on M and h. For example, given $Z \in L$, set $X(t) = h(Z,t) \in M^o$ for $t \in (0,1]$, and let R(t) be the curvature tensor of M^o at X(t); then it is required that $\|R(t)\| = O(t^{-2})$, independently of Z. The methods of this paper generally use order-of-magnitude estimates like this one to prove the convergence as $t \to 0+$ of various properties of $\exp_{X(t)}$, thus establishing them also at P.

AMS(MOS) subject classifications (1970): Primary 53 B 20; Secondary 32 C 40, 32 B 15.

CHAPTER 1: INTRODUCTION

Many fundamental concepts of the differential geometry of smooth manifolds have been extended to manifolds with singularities; for example in the work of Sullivan [14] on differential forms in simplicial complexes; in that of Cheeger on Hodge theory [3]; in that of Griffiths [5] on the limits of curvature integrals on non-singular analytic varieties as they converge to a singular one; in the work of A. D. Aleksandroff and his school on curvature in 2-manifolds with quite general metric [1]; and in my own work [13] on geodesics in piecewise linear manifolds. Less attention has been given to generalizing to manifolds with singularities the exponential map and the formulae for the first and second variation of arc length. Such a program would permit the use of the calculus of variations to study geodesics in manifolds with singularities.

The present paper shows that, under certain hypotheses on the intrinsic geometry of a metric space, there exists an exponential map at a singular point. For example let M be a complex analytic hypersurface $\{f = 0\}$ in $\mathbb{C}^{n+1}$ with an isolated singular point P. It is shown that when f is of a certain type, there does exist an exponential map at P, whereas an example is given of another f for which no such map can exist. In order to state the results more precisely, let me first give a definition of an "exponential map" sufficiently general for this paper.

Let M be a stratified space, and assume the strata M_i of M are equipped with compatible C^∞ Riemannian metrics. Any affine complex analytic variety $M \subseteq \mathbb{C}^{n+1}$ can be expressed in this form (see Whitney [19]); each M_i has the metric induced on it as a C^∞ submanifold of $\mathbb{C}^{n+1}$. Then one can define "piecewise C^1" paths in M, and the "length" and "energy" of such paths. In this discussion every path β will be assumed to have domain $[0,1]$ and will be said to go "from $\beta(0)$ to $\beta(1)$." A metric is defined on

Received by the editors November 11, 1980.

Research supported in part by N.S.F. grant No. MCS-7802147 and in part by a grant from the City University of New York PSC-BHE Research Award Program.

M by the rule

$$d^M(X,Y) = \inf\{\text{length }(\beta)\}$$

where β ranges over all piecewise C^1 paths from X to Y. β is a <u>geodesic</u> if it has least energy among all piecewise C^1 paths from $\beta(0)$ to $\beta(1)$; then β also has shortest length.

Now let $P \in M$. I say that M <u>has an exponential map at</u> P, or that $\exp_P$ <u>exists</u>, if (1.1) - (1.5) hold.

(1.1) There exists a neighbourhood U of P in M such that for every $X \in U$, there is a geodesic γ_X from P to X with $\mathrm{im}(\gamma_X) \subseteq U$.

(1.2) Given $X \in U$, then γ_X is unique.

The third condition says that the γ_X for $X \in U$ can be distinguished by their "tangent vectors" at P. For this one must first single out a class of piecewise C^1 paths from P, called <u>paths</u> C^1 <u>at</u> P, of which it is reasonable to say that they have tangent vectors at P; this class must include all the geodesics γ_X. Paths α and β which are C^1 at P are called <u>equivalent</u> if

$$d^M(\alpha\sigma,\beta\sigma) = o(\sigma).$$

Let $\hat{\alpha}$ denote the equivalence class, or tangent vector at P, of α, and let $\hat{P}$ denote the class of the constant path at P. Set $T_PM = \{\hat{\alpha}\}$. Then T_PM, together with a certain natural metric (see (2.25)), is called the <u>tangent metric space of</u> M <u>at</u> P. Assuming that (1.1) and (1.2) hold, there is a map $\exp_P^{-1}$: $U \to T_PM$ defined by $\exp_P^{-1}(X) = \hat{\gamma}_X$. I require:

(1.3) $\exp_P^{-1}$ is one-to-one.

(1.4) $\exp_P^{-1}(U)$ is a neighbourhood V of P in T_PM.

Now $\exp_P^{-1}$ has an inverse function $\exp_P$: $V \to U$ between neighbourhoods of P in T_PM and P in M. It is required that $\exp_P$ be a homeomorphism:

(1.5) $\exp_P^{-1}$ and $\exp_P$ are continuous.

For example, let M be a complex analytic variety defined by some homogenous polynomials in $\mathbb{C}^{n+1}$, so that M is a cone from the origin, P. Then $T_PM = M$ and $\exp_P$ is the identity map. This is true more generally if

M is any metric cone (see (1.16)).

(1.6) The main theorem of this work is a partial generalization of the last example. It states that if M is "approximately" a cone from P on a closed, compact manifold, then $\exp_P$ exists. The requisite notion of approximation now follows; for the notation see (1.10) ff.

P is defined to be an <u>isolated conical singularity of</u> M if:

(1.6.1) $M^o = M-\{P\}$ is a Riemannian n-manifold without boundary of class C^∞, and if, in addition, there exist:

(1.6.2) a closed, compact (n-1)-manifold L of class C^∞, and

(1.6.3) a homeomorphism, for some $\underline{t} > 0$, $h\colon c^{\underline{t}}L \to M$, called a <u>chart for</u> M <u>at</u> P, onto a neighbourhood of P in M, such that:

I $h(c) = P$;

II $h\colon (c^{\underline{t}}L)^o \to M^o$ is a C^∞ diffeomorphism;

III for each $Z \in L$, $\lim_{t\to 0^+} |Dh(Z,t)\cdot\partial_{\underline{t}}| = 1$;

IV whenever $t > 0$ and $u \in T_{(Z,t)}(L \times t)$, then $\langle Dh\cdot u, Dh\cdot\partial_{\underline{t}}\rangle = 0$;

V there exists $k > 0$ such that whenever $t > 0$ and $\psi \in T_{h(Z,t)}M^o$, then $|D^M_\psi(Dh\cdot\partial_{\underline{t}}) - t^{-1}(Dh\cdot\partial_{\underline{t}})^\perp(\psi)| = |\psi|O(t^{-1+2k})$;

VI there exists $K > 0$ such that the sectional curvature of 2-planes on M^o is everywhere $<K$;

VII let $\psi, \omega \in T_{h(Z,t)}M^o$; then with $k > 0$ as in V,

(i) $\|R^M(\psi,\omega)\| = |\psi \vee \omega|O(t^{-2})$,

(ii) $\|R^M(Dh\cdot\partial_{\underline{t}},\omega)\| = |\omega|O(t^{-2+k})$,

(iii) $|R^M(Dh\cdot\partial_{\underline{t}},\omega)(Dh\cdot\partial_{\underline{t}})| = |\omega|O(t^{-2+2k})$.

The main theorem can now be stated:

THEOREM 1.7. Let P be an isolated conical singularity of M. Then there exists an exponential map of M at P satisfying (1.1) - (1.5). (See Theorem 6.9.)

(1.7.1) REMARK. In (1.6.3, VI) the requirement $K > 0$ is only for convenience of exposition: any upper bound on the sectional curvature of M^o can be made to serve the purpose. An example of Aleksandroff [1, Kap. I, §10] shows that some sort of upper bound on the curvature is necessary.

(1.7.2) REMARK. Let M be in fact non-singular at P as a C^∞ Riemannian manifold. Let L be the unit sphere in T_PM, so $c^{\underline{t}}L$ is the closed ball of radius $\underline{t}$ about P in T_PM. Then if $\underline{t} > 0$ is small enough, $h = \exp_P : c^{\underline{t}}L \to M$ satisfies (1.6.3). On the other hand let L satisfy (1.6.2) and let M be the metric cone $c^\infty L$; then the identity map is an exponential map at c; yet (1.6.3, VI) need not hold. All the other hypotheses of (1.6.3) do hold for $c^\infty L$, and (1.6.3, VII) provides an order-of-magnitude upper bound on the sectional curvature; so it may be possible to prove Theorem 1.7 without assuming (1.6.3, VI). In (7.25) I conjecture that an exponential map exists in a certain type of example where (1.6.3, VII) fails.

Now let M be a complex analytic hypersurface in $\mathbb{C}^{n+1}$ with equation $f = 0$ and let P be an isolated singular point of M in the sense of Milnor [9]. I say that P is unbranched if it is also isolated as a critical point of the initial homogeneous polynomial of f. The main applications of Theorem 1.7 are the next two theorems.

THEOREM 1.8. Let P be an unbranched isolated singular point of a complex analytic hypersurface $M \subseteq \mathbb{C}^{n+1}$, and let M^o be given the Riemannian metric on $\mathbb{C}^{n+1}$. Then T_PM is homeomorphic to M near P, and every geodesic γ from P is C^1 at P, so $\hat{\gamma}$ exists in T_PM. If also (1.6.3,VI) holds in M^o near P, then P is an isolated conical singularity of M; hence M has an exponential map at P. (See Corollaries 7.6.1 and 7.6.2.)

THEOREM 1.9. Let P be any singular point of a complex analytic curve $M \subseteq \mathbb{C}^2$ and let M^o be given the metric of Theorem 1.8. The P is an isolated conical singularity of M; so M has an exponential map at P. (See Theorem 7.11.)

The polynomial $f(x,y,z) = x(y^3 - x^2) - z^2$ gives rise to a surface $M \subseteq \mathbb{C}^3$; the origin P, though an isolated singular point, is not unbranched; and M cannot have an exponential map at P. (See Example 7.23.) I might point out the difference in this regard between weighted homogeneous polynomials, such as f, and homogeneous polynomials.

Outline of the Paper

Chapter 2 studies the "first-order" properties of a chart h: $c^{\underline{t}}L \to M$ at P; that is, those which do not depend upon the curvature hypotheses VI and VII of (1.6.3). This is enough to prove results about individual geodesics from P, such as (1.1) (Corollary 2.17.4). The general method is to compare how geodesics from P in M and the images under h of generating rays of $c^{\underline{t}}L$ approximate each other. The tangent metric space T_PM is defined in (2.25). The definitions initially depend on the choice of a chart h. A path β from P in M is called "C^1 at P with respect to h" if, on some initial segment $[0,\mu]$ such that $\mathrm{im}(\beta\restriction[0,\mu]) \subseteq \mathrm{im}(h)$,

$$\beta:(0,\mu] \to M^{o} \text{ is } C^1,$$

$$|\beta'(\sigma)| \text{ is bounded on } (0,\mu],$$

and, writing $h^{-1}\circ\beta(\sigma) = (Z(\sigma),t(\sigma))$,

$$\lim_{\sigma\to 0^+} (Z(\sigma),t(\sigma)/\sigma) \text{ exists in } c^{\infty}L.$$

(See (2.11).) Paths β_1 and β_2, both C^1 at P with respect to h, are "equivalent" if the corresponding limits are equal, and T_PM is the set of equivalence classes. Proposition 2.14 implies that T_PM is independent of the choice of h, and Proposition 2.13 that every geodesic from P is C^1 at P with respect to any chart at P. There is a natural metric d^T on T_PM (see (2.25.1)). Theorem 2.26 shows that in this metric, T_PM is an infinite cone and that any chart h can be "differentiated at P" to yield a homeomorphism T_Ph: $c^{\infty}L \to T_PM$. Thus

$$e(h)_P = h\circ(T_Ph)^{-1}: \quad T_PM \to M$$

is a (non-canonical) homeomorphism near P.

In Chapter 3 the curvature hypothesis (1.6.3, VI) is used to prove (1.2) (see Theorem 3.3). An important step in the proof is an estimate of the order of magnitude of the radius of convexity of M^{o} at X (Proposition 3.7). Let γ_X and γ_* be geodesics from P to a nearby point X. Proposition 3.7 and results from Chapter 2 show that for each λ there is a unique geodesic β^{λ} from $\gamma_X(\lambda)$ to $\gamma_*(\lambda)$; and by uniqueness, β^{λ} must vary continuously in λ. I now apply Rauch's comparison theorem to M^{o} and an n-sphere

S of radius $K^{-\frac{1}{2}}$, whose sectional curvature is everywhere greater than that of M^o by (1.6.3, VI). It follows that γ_X and γ_*, regarded as geodesics from X in M^o, must diverge -- as measured by the length of β^λ -- faster than do corresponding geodesics from a point of S. Therefore γ_X and γ_* can converge at P only if $\gamma_X = \gamma_*$, which proves Theorem 3.3. As mentioned above, the theorem implies the existence of an "inverse-exponential" map e_P^{-1}: $M \to T_P M$ near P.

Chapter 4 shows that e_P^{-1} is "infinitesimally one-to-one. Let γ_X be the geodesic from P to X of (1.2) and let $\psi \in T_P M$. Then there exists a unique Jacobi field $\underline{J}\{\psi\}$ along γ_X with final value $\underline{J}\{\psi\}(\gamma_X 1) = \psi$ and "initial value $\vec{0}$" in the sense that

$$\lim_{\lambda \to 0^+} \underline{J}\{\psi\}(\gamma_X \lambda) = \vec{0}$$

(Theorem 4.13). The proof uses a boundary-value form of Rauch's comparison theorem to compare M to S. This gives an upper bound on $|\underline{J}(\gamma\lambda)|$ in terms of $|\underline{J}(\gamma 1)|$ whenever $\underline{J}$ is a Jacobi field along a (fairly short) geodesic γ in M^o such that $\underline{J}(\gamma 0) = 0$. In particular let $\underline{J}_\nu\{\psi\}$ be the Jacobi field along $\gamma_X \restriction [\nu, 1]$ such that $\underline{J}_\nu\{\psi\}(\gamma_X 1) = \psi$ and $\underline{J}_\nu\{\psi\}(\gamma_X \nu) = \vec{0}$; then it follows that $\lim_{\nu \to 0^+} \underline{J}_\nu\{\psi\}$ exists and is the required $\underline{J}\{\psi\}$. The other main result of the chapter is that $\underline{J}\{\psi\}$ can to some extent be "differentiated" at P; at least

$$\lim_{\lambda \to 0^+} |\underline{J}\{\psi\}(\gamma_X \lambda)|/\lambda \text{ exists;}$$

moreover this limit is non-zero whenever $\psi \neq 0$ (Proposition 4.21). This last result is the sense in which e_P^{-1} is "infinitesimally" one-to-one. The proof of the proposition requires a lower bound on $|\underline{J}\{\psi\}(\gamma_X \lambda)|$ in terms of $|\psi|$. Again Rauch's comparison theorem, in boundary-value form, is used. Since (1.6.3, VII) provides a lower estimate on the curvature of M along γ_X, one needs a comparison space M, with a distinguished point P, whose curvature along geodesics g from P equals this lower estimate, and whose Jacobi fields along g with "initial value $\vec{0}$" can be computed. Such a model is constructed in (4.4).

Chapter 5 deals with one-parameter families of geodesics from P. Let

β be a C^∞ path in M^o such that $\beta(0) = X$ and $\beta'(0) = \psi$. Then (Theorem 5.13) for all $\lambda \in (0,1]$,

$$\underline{\partial}_\sigma \gamma_{\beta\sigma}(\lambda)\big|_{\sigma=0} \text{ exists and equals } \underline{J}\{\psi\}(\lambda)$$

(as occurs in the non-singular situation). The method of proof is to show first that if β is short enough and $\nu > 0$ small enough, then there exists a one-parameter family of weak geodesics $\gamma_{\nu,\sigma}$ from $\gamma_X(\nu)$ to $\beta(\sigma)$ (Proposition 5.2). Now

$$\underline{\partial}_\sigma \gamma_{\nu,\sigma}(\lambda)\big|_{\sigma=0} = \underline{J}_\nu\{\psi\}(\lambda),$$

so in view of Theorem 4.13 it remains to be shown that

$$\lim_{\nu\to 0^+} \gamma_{\nu,\sigma}(\lambda) = \gamma_{\beta\sigma}(\lambda)$$

(Proposition 5.9), and that

$$\underline{\partial}_\sigma \gamma_{\beta\sigma}(\lambda)\big|_{\sigma=0} \text{ exists and equals } \lim_{\nu\to 0^+} \underline{\partial}_\sigma \gamma_{\nu,\sigma}(\lambda)\big|_{\sigma=0}$$

(Lemma 5.10).

The first half of Chapter 6 leads up to Theorem 6.9, which proves Theorem 1.7. First the proof of (1.3) is completed, that e_P^{-1} is one-to-one (Proposition 6.4); (1.5) is a by-product of the argument. There is a twist in the proof of the proposition. Let the path β above go from X to Y, and set $\beta^\lambda(\sigma) = \gamma_{\beta\sigma}(\lambda)$; then β^λ is a path from $\gamma_X(\lambda)$ to $\gamma_Y(\lambda)$, and by Theorem 5.13, β^λ is differentiable: $\beta^{\lambda\prime}(\sigma) = \underline{J}\{\beta'\sigma\}(\lambda)$. But β^λ need not be C^1, so estimates such as

$$d^M(\gamma_X\lambda, \gamma_Y\lambda) \le \text{"length of } \beta^\lambda\text{"} = \text{"}\int_0^1 |\beta^{\lambda\prime}(\sigma)|\,d\sigma\text{"}$$

must be treated more carefully. I have therefore had to discuss intrinsic metrics at some length at the beginning of Chapter 2, in particular the connection between the length of a C^1 path, $\int |\beta'(\sigma)|\,d\sigma$, and the length of a continuous path as defined by Aleksandroff (see (2.5)). To prove (1.4), that e_P^{-1} is surjective near P, I show essentially that e_P^{-1} is pointwise so close to $e(h)_P^{-1}$ near P that these two maps are homotopic as maps of pairs $(M,M^o) \to (T_PM,(T_PM)^o)$; since $e(h)_P^{-1}$ is surjective, the result follows.

The second half of Chapter 6 analyses T_PM. The main results are that e_P: $T_PM \to M$ near P satisfies all the axioms of (1.6) on a chart except (1.6.2), which refers to the differentiability structure of $(T_PM)^o$ (Theorem 6.13); and (Theorem 6.15) that the metric d^T is intrinsic in the sense of Aleksandroff (see (2.5)).

Chapter 7 contains the applications to complex analytic hypersurfaces, Theorems 1.8 and 1.9. The order of magnitude estimates are not very delicate in the first theorem: one may use $k = 1/2$ in V and VII of (1.6.3). But k must be chosen more carefully in the second theorem. Each branch of M at P can be described as

$$\xi(x) - y^p = 0, \text{ where } p < q = \text{order}_P(\xi);$$

then I need $k = (1/2)\min(1,(q/p) - 1)$ (see Theorem 7.11). In the context of Theorem 1.8, T_PM is homeomorphic to the Whitney tangent cone of M at P (Theorem 7.10). In the context of Theorem 1.9, T_PM is more subtle, but its metric structure can be completely described. M is the one-point union at P of its branches M_i, and the Whitney tangent cone T of M at P is the union of corresponding lines T_i (any two of which either coincide or meet only in P). Then T_PM is the one-point union at P of the T_PM_i. For each i there is a natural map $\hat{\pi}$: $T_PM_i \to T_i$ which is a p_i-fold covering map branched at P, where p_i is the multiplicity of M_i. Moreover $\hat{\pi}$: $(T_PM_i)^o \to (T_i)^o$ is a local isometry. (See Theorem 7.18.) The theorem thus interprets geometrically the fact that the algebraic tangent cone of M at P is the union of the T_i, each taken with multiplicity p_i.

Finally some examples and conjectures indicate the limitations on and possibilities of generalizing Theorems 1.7, 1.8 and 1.9.

Notations and Conventions

(1.10) Topology. Let N be a topological space, $Q \in N$ a distinguished point. Then

(1.10.1) N^o denotes $N - \{Q\}$.

Q may be distinguished only implicitly; for example Q may be an isolated singular point of N, or the origin of a vector space.

Let N* be a second topological space and let a neighbourhood U of Q in N and a map $f: U \to N^*$ be given. I shall often suppress mention of U and say simply that a map

(1.10.2) $f: N \to N^*$ near Q

is given.

(1.11) <u>Metric geometry</u>. Let N be a metric space with metric d^N.

Then for any $r > 0$,

(1.11.1) $B(X,N;r) = \{Y \in N \text{ such that } d^N(X,Y) < r\}$; and

(1.11.2) $B^0(X,N;r) = B(X,N;r) - \{X\}$.

(1.12) <u>Orders of magnitude</u>. Let x be a real variable and f a real-valued function depending on x and possibly on other variables.

The notations

(1.12.1) $f = O(x^k)$ and $f = o(x^k)$

will imply that f converges to zero with x as described, <u>uniformly</u> in all other variables. When $k = 0$, (1.12.1) means, respectively, that f is bounded in x or that $f \to 0$ as $x \to 0$, again uniformly in all ohter variables. When restrictions must be placed on other variables, they will almost always be expressed in the notation; for example

(1.12.2) $xy = O(x; \text{fixed } y)$.

But if, in this example, an upper bound on $|y|$ has been long in force (not just in the argument at hand), then I shall often write simply: $xy = O(x)$.

(1.13) <u>Linear algebra</u>. Let V be a real or complex vector space and let $v,w \in V$. According to the context, V will always have either

(1.13.1) a positive-definite inner product $\langle\ ,\ \rangle$, if V is a real vector space,

or

(1.13.2) a positive-definite Hermitian product $\underline{\langle}\ ,\underline{\rangle}$, if V is a complex vector space.

In the latter case V has the associated inner product

(1.13.3) $\langle v,w \rangle = \operatorname{Re} \underline{\langle} v,w \underline{\rangle} = (1/2)(\underline{\langle} v,w \underline{\rangle} + \underline{\langle} w,v \underline{\rangle})$.

Either product (1.13.1) or (1.13.2) gives rise to a norm $|\ |$ on V. Since the vector spaces used in this work will usually be tangent spaces rather than cotangent spaces, I use the notation

(1.13.4) $v \vee w$ for the Grassmann product of v and w

(compare Whitney [18, Introduction, §9] for example).

When $v \neq 0$ I use the following notation for the components of w respectively parallel to and orthogonal to v:

(1.13.5) $v^{\parallel}(w) = |v|^{-2}\langle w,v \rangle v$, and $v^{\perp}(w) = w - v^{\parallel}(w)$,

in case V is real; and

(1.13.6) $v^{\underline{\parallel}}(w) = |v|^{-2}\underline{\langle} w,v \underline{\rangle} v$, and $v^{\perp}(w) = w - v^{\underline{\parallel}}(w)$,

in case V is complex. Using this notation,

(1.13.7) $$|v \vee w| = \begin{cases} 0 & \text{, if } v = \vec{0}, \\ |v|\,|v^{\perp}(w)| & \text{, if } v \neq \vec{0}. \end{cases}$$

When $V = \mathbf{C}^{n+1}$ I shall only use the standard Hermitian product

(1.13.8) $$\underline{\langle} v,w \underline{\rangle} = \sum_{i=1}^{n+1} v_i \bar{w}_i,$$

where $v = (v_1,\ldots,v_{n+1})$ and $w = (w_1,\ldots,w_{n+1})$. In working with components of vectors, the Kroneker symbol on a given set of indices is helpful:

(1.13.9) $\delta^i_j = \begin{cases} 1, & \text{if } i = j; \\ 0, & \text{otherwise.} \end{cases}$

Let D: $V \to V^*$ be a linear transformation and let $v \in V$. I write

(1.13.1) $D \cdot v$ for the value of D on v.

The norm of D is

(1.13.11) $\|D\| = \sup\{|D \cdot v|/|v| \text{ for } v \quad V^o\}$

(see (1.10.1)).

(1.14) Calculus. The manifolds used in this work will have class C^∞ unless otherwise specified. The tangent space of N at X and the tangent bundle of N will be denoted respectively

(1.14.1) $T_X N$ and $T(N)$.

I shall often write $T_X N$ instead of $T_X(N^o)$ even when N has a singular point $P \neq X$.

The derivative of a differentiable function f: $N \to N^*$ will be denoted

(1.14.2) $Df\colon\ T(N) \to T(N^*)$,

and the derivative of f at X will be written

(1.14.3) $Df(X)\colon\ T_X N \to T_{fX} N^*$.

Given $v \in T_X N$, the value of Df(X) on v will be called the derivative of f in the direction v; it is denoted, by (1.13.10),

(1.14.4) $Df(X) \cdot v$, abbreviated sometimes to $Df \cdot v$ or $\partial_v f$.

To denote vector fields on manifolds I shall use underlined symbols such as

(1.14.5) $\underline{v}\colon\ N \to T(N)$.

When regarding $\underline{v}$ as an operator on functions, I shall write

(1.14.6) $\partial_{\underline{v}}(f)\colon\ N \to T(N^*)$, defined by

$$(\partial_{\underline{v}} f)(X) = \partial_{\underline{v}(X)} f = Df(X)\cdot\underline{v}(X) = (Df\cdot\underline{v})(X),$$

in the notation of (1.14.4).

Given local coordinates $(x_1,\ldots,x_n)$ on N, the coordinate vector fields will be denoted

(1.14.7) $\underline{\partial}_{x_i}$ or $\underline{\partial}_i$, short for $\frac{\partial}{\partial x_i}$.

In case N is an affine space (over $\mathbb{R}$ or $\mathbb{C}$) and N* a vector space, the second derivative of f, if it has one, will be written

(1.14.8) $v\cdot D^2 f(X)\cdot w$.

(1.15) Riemannian geometry. Let N be a Riemannian manifold. Its metric will be denoted

(1.15.1) ds^N, $\langle\ ,\ \rangle^N$, or g^N.

In terms of a local coordinate system $(x_1,\ldots,x_n)$ and (1.14.7),

(1.15.2) $g^N_{ij}(X) = \langle\underline{\partial}_i,\underline{\partial}_j\rangle^N(X)$.

As in (1.14.1) I shall write g^N instead of g^{N^o} even when N has a singular point. I shall also write $g^{\mathbb{C}}$ instead of $g^{\mathbb{C}^{n+1}}$. The same abbreviations will apply to all the notation of (1.15).

The metric d^N on a Riemannian manifold will always be the one associated to its Riemannian metric as in (2.4) or in Gromoll, Klingenberg and Meyer (henceforth: GKM) [6, §5.1].

The Levi-Civita connection associated to g^N will be denoted D^N; its components Γ^N in a given coordinate system are determined by

(1.15.3) $D^N_{\underline{\partial}_k}(\underline{\partial}_j) = \Sigma_i \Gamma^{Ni}_{jk}\underline{\partial}_i$.

The curvature tensor of D^N will be denoted R^N. Let $\underline{v}$ and $\underline{w}$ be vector fields on N with values $\underline{v}(X) = v$, $\underline{w}(X) = w$; then

(1.15.4) $R^N(v,w) = (D^N_{\underline{v}}D^N_{\underline{w}} - D^N_{\underline{w}}D^N_{\underline{v}} - D^N_{[\underline{v},\underline{w}]})(X)$.

In local coordinates R^N has components

(1.15.5) $R^N_{hkij}(X) = \langle R^N(\underline{\partial}_h,\underline{\partial}_k)\underline{\partial}_i,\ \underline{\partial}_j\rangle(X)$.

The sectional curvature K^N is defined on a 2-plane $\Pi \subseteq T_X N$ in terms of a basis (v,w) of Π by

(1.15.6) $K^N(\Pi) = \langle R^N(v,w)v, w\rangle / |v \wedge w|^2$

(see (1.13.7)).

(1.15.7) Caution. By a weak geodesic in N I shall mean what is traditionally called a "geodesic" in differential geometry: a C^2 path β that satisfies the second-order differential equation $(D^N_{\beta',\sigma}\beta')(\sigma) \equiv 0$. The term "geodesic," as used in the present work, is defined in (2.7); in the context of Riemannian manifolds it refers to a shortest weak geodesic.

The notions of geodesic and weak geodesic coincide locally. It is known that given $X \in N$, then for all sufficiently small $r > 0$,

(1.15.8) whenever $Y_1, Y_2 \in B(X,N;r)$ (of (1.11.1)), there is a unique weak geodesic β in $B(X,N;r)$ from Y_1 to Y_2 with given domain; further, β is the unique shortest weak geodesic from Y_1 to Y_2 in N with the given domain.

This follows from GKM [6, §5.2] by taking r less than half the radius of convexity of N at X (which is defined there).

The exponential map of N at X will be denoted

(1.15.9) $\exp_X$ or simply e_X.

The map $e_X\colon T_X N \to N$ near X (see (1.10.2)) is defined by $e_X(v) = \beta(1)$, where β is the weak geodesic with domain $[0,1]$ determined by the initial conditions $\beta(0) = X$, $\beta'(0) = v$.

(1.16) Cones. Let L be a compact topological space. The cone on L with vertex c and height $\underline{t}$ is defined, for $\underline{t} \in (0,\infty]$, by

(1.16.1) $c^{\underline{t}}L = L \times [0,\underline{t}]/L \times 0$, for $\underline{t} \in (0,\infty)$;

$c^{\infty}L = L \times [0,\infty)/L \times 0$;

$c^1 L$ is abbreviated to cL.

In each case the cone is given the identification topology. The base of $c^{\underline{t}}L$ is $L \times \underline{t}$; this is identified with L only when $\underline{t} = 1$. Whenever

$t \in (0,\underline{t}]$ there is a natural inclusion

(1.16.2) $i_t\colon\ L \to c^{\underline{t}}L$ defined by $i_t(Z) = (Z,t)$.

A map $f\colon\ c^{\underline{t}}L \to c^{\underline{t}^*}L^*$ is <u>conical</u> if there exists a map $f'\colon\ L \to L^*$ such that

(1.16.3) $f(Z,t) = (f'Z,t)$.

The multiplicative group $\mathbb{R}_+ = \{r \text{ such that } r > 0\}$ acts on $c^{\infty}L$ by

(1.16.4) $r(Z,t) = (Z,rt)$;

this map will be called the <u>dilation with factor</u> r. I shall use the same notation to describe $r\colon\ c^{\underline{t}}L \to c^{\underline{t}^*}L$ whenever $\underline{t}^* \geq r\underline{t}$.

When L is a C^{∞} manifold, I use the abbreviation

(1.16.5) $\underline{\partial}_t = \dfrac{\partial}{\partial t}$

for the unit radial vector field on $(c^{\underline{t}}L)^{\circ}$, as in (1.14.7).

Given a metric d^L on L, a metric $d^{c^{\underline{t}}L}$ is defined on $c^{\underline{t}}L$ by

(1.16.6) $$d^{c^{\underline{t}}L}((Z_1,t_1),(Z_2,t_2)) =$$
$$= \min\{t_1+t_2, [t_1{}^2+t_2{}^2-2t_1t_2 \cos(d^L(Z_1,Z_2))]^{1/2}\}.$$

If L is a Riemannian manifold, then so is $(c^{\underline{t}}L)^{\circ}$ under the metric

(1.16.7) $(ds^{c^{\underline{t}}L})^2(Z,t) = t^2(ds^L)^2(Z) + dt^2$.

This is consistent with the convention of (1.15): if d^L is the metric associated to ds^L, then the metric $d^{c^{\underline{t}}L}$ of (1.16.6) is also associated to $ds^{c^{\underline{t}}L}$ of (1.16.7). When either (1.16.6) or (1.16.7) holds, $c^{\underline{t}}L$ is called a <u>metric cone</u>. An equivalent definition can be given in terms of dilations: corresponding to (1.16.6) is

(1.16.8) $d^{c^{\underline{t}}L}(r(Z_1,t_1),r(Z_2,t_2)) = r d^{c^{\underline{t}}L}((Z_1,t_1),(Z_2,t_2))$,

whenever $r \in \mathbb{R}_+$ and $(Z_i,t_i) \in c^{\underline{t}}L$; while corresponding to (1.16.7) is

(1.16.9) $(Dr\cdot ds^{c^{\underline{t}}L})(Z,t) = r\, ds^{c^{\underline{t}}L}(r(Z,t))$, whenever $r \in \mathbb{R}_+$ and $(Z,t) \in (c^{\underline{t}}L)^o$ (see (1.14.4)).

(1.16.10) Remark. The parameter $\underline{t}$, as used in (1.6.3), permits the construction of geometrically natural charts in Chapters 4 and 7. Elsewhere, to simplify the notation, I shall usually assume $\underline{t} = 1$ and write $h\colon cL \to M$ for a chart at P. Given a chart $h^*\colon\ c^{\underline{t}}L \to M$ one can construct such an h: by restriction, if $\underline{t} \geq 1$; while if $\underline{t} \in (0,1)$, let $f\colon\ [0,1] \to [0,\underline{t}]$ be a C^∞ diffeomorphism which is the identity near 0, and set $h(Z,t) = h^*(Z,ft)$.

(1.17) Some useful estimates.

(1.17.1) $(\sin\theta)/\theta > 1 - \theta^2/6 > 1/2$ on $(0,\pi/2]$.

(1.17.2) $\theta/(\sin\theta) < 1 + \theta^2/6 < 3/2$ on $(0,\pi/2]$.

It follows from these two estimates and the fact that $\sin(\theta) < \theta$ on $(0,\pi/2]$ that

(1.17.3) $(1/2)(\theta_1/\theta_2) < (\sin\theta_1)/(\sin\theta_2) < (3/2)(\theta_1/\theta_2)$ for $\theta_1,\theta_2 \in (0,\pi/2]$.

Let $a,b > 0$ and $c \geq 0$. Then

(1.17.4) $$a,b\begin{cases} \leq (a+c)/(b+c) & \text{if } a \leq b, \\ \geq (a+c)/(b+c) & \text{if } a \geq b; \end{cases}$$

with equality only if $a = b$ or if $c = 0$.

(1.18) Typographical conventions

(1.18.1) All theorems, propositions, lemmas, corollaries, definitions, displayed equations, remarks and examples are numbered consecutively. Thus in Chapter A, Theorem A.B.D is the Dth assertion of Theorem A.B; (A.B.E) may be an estimate derived in the proof of the assertion to which there will be later occasion to refer; and Corollary A.B.F is a direct consequence of Theorem A.B.

(1.18.2) The only exception to (1.18.1) is that during the proof of a proposition which has several assertions I shall refer to these assertions simply by their last numbers. For example, in the proof of Theorem A.B.D, the phrase "by (C)" will mean "by Theorem A.B.C."

(1.18.3) For simplicity's sake I shall often -- but not very consistently -- omit the sort of parentheses that denote functional dependence; thus $\beta_1(\sigma)$, but $d^M(\beta_1\sigma,\beta_2\sigma)$.

(1.18.4) Letters X,Y and Z denote points; P always refers to a singular point.

Script capitals $\mathcal{A}$, $\mathcal{B}$, $\mathcal{H}$, $\mathcal{K}$ are used for numerical bounds in estimates. $\mathcal{E}$ and $\mathcal{L}$ denote energy and length respectively. See also (1.18.7).

λ, μ, ν, σ, τ, r, s, t are real-valued parameters, often denoting position along a path. x, y and z are real or complex coordinates. τ and t are used in connection with the notion of distance from a singular point; see (1.18.5) and (1.18.6).

I shall try to use other symbols systematically as follows.

(1.18.5) In M I use Greek letters. Thus paths in M are denoted α,β and γ; γ usually refers to a geodesic from P, and τ denotes distance from P. Vectors in T(M) are denoted ψ, ω.

(1.18.6) Let h: cL $\to$ M be a chart (compare (1.16.10)). Points of cL are denoted (Z,t). Paths in cL are labelled a, b and g; they often correspond via h to α, β and γ respectively. A path a always is a geodesic from c, so a(s) = (Z,st). Note that the coordinate t on cL represents the distance from c in the metrics (1.16.7) and (1.16.8).

(1.18.7) In spaces $\mathcal{S}$ or $\mathcal{M}$, with which M is being compared, I use script letters. For example, g will denote a geodesic from a point $\mathcal{P}$ corresponding to P.

(1.18.8) Given $X \in M^o$, I use a tilde to denote entities in T_XM. Usually symbols differing only by a tilde are related by e_X (of (1.15.9)); for example $e_X(Y) = \tilde{Y}$, $e_X \circ \beta = \tilde{\beta}$.

(1.18.9) A circumflex will denote entities in T_PM. Again, symbols differing only by a circumflex will usually correspond under e_P^{-1} or e_P. For example, the equivalence class of a path β from P in M (when defined) is

$\hat{\beta} \in T_p M$ (see (2.25)); and the map e_p will eventually be defined by: given $\hat{X} \in TPM$, then $e_p(X)$ is that $X \in M$ such that $\hat{X} = \gamma_{\hat{X}}$.

CHAPTER 2: FIRST-ORDER PROPERTIES

Introduction

In this section the metric tangent space $T_P M$ of M at P is defined. Since a tangent space is a "first-order" construct one would expect the "second-order" curvature hypotheses VI and VII of (1.6.3) not to be needed. Such is the case: in this chapter only (1.6.1), (1.6.2) and I--V of (1.6.3) are assumed. To define $T_P M$ one must first select a class of curves from P which have "tangent vectors" at P. In case M is a cone cL, I take the class of C^1 paths b from c (see (2.1)) which are "C^1 at c" in the sense that, writing $b(\sigma) = (Z(\sigma), t(\sigma))$,

$$\lim_{\sigma \to 0^+} (Z(\sigma), t(\sigma)/\sigma) \text{ exists in } c^\infty L$$

(see (2.7)). For general M, let h: $cL \to M$ be a chart at P; then a path β from P is "C^1 at P" if $\beta = h \circ b$, where b is C^1 at c. Two of the main goals of this chapter are to prove that this class of paths includes all geodesics from P (Proposition 2.9) and that it is independent of the choice of chart at P (Proposition 2.10). I shall now outline the proof of Proposition 2.9.

Let γ be a geodesic from P, which I assume parametrized by arc length to simplify this discussion. Set $g = h^{-1} \circ \gamma$; it is required to show that g is C^1 at c. Writing $g(\tau) = (Z(\tau), t(\tau))$, this can be expressed as requiring that $\lim_{\tau \to 0^+} Z(\tau)$ and $\lim_{\tau \to 0^+} t(\tau)/\tau$ exist. It is relatively easy to show that $\lim_{\tau \to 0^+} t(\tau)/\tau = 1$. First, Dh and Dh^{-1} are bounded in norm (Lemma 2.12.1); this uses (1.6.3,V). Now $t(\tau)$ is the length of an integral curve of $Dh \cdot \underline{\partial}_t$ from P to $\gamma(\tau)$; so $t(\tau) \geq$ length of $\gamma \restriction [0,\tau]$, namely τ. On the other hand, $\frac{d}{d\tau} t(\tau)$ is the size of the components of $Dh \cdot \underline{\partial}_t$ in the direction of $\gamma'(\tau)$; using (1.6.3,IV) it follows that $|Dh \cdot \underline{\partial}_t|(\gamma\tau) \approx 1$ (Lemma 2.12.2). This gives an estimate of $\frac{d}{d\tau} t(\tau)$ which, when integrated, implies that $t(\tau)$ cannot be much more than τ. Thus $t(\tau) \approx \tau$; and when the estimates are made more carefully (Lemma 2.13.1) the required limit $t(\tau)/\tau \to 1$ follows. The proof that $\lim Z(\tau)$ exists is more delicate. The orders of magnitude involved are worse than before because

$$\left|\frac{d}{d\tau}Z(\tau)\right| = |\partial_{\underline{t}}^{\perp}(g'\tau)|/t(\tau),$$

where $\partial_{\underline{t}}^{\perp}(g'\tau)$ is as in (1.16.5) and (1.13.5). By Lemma 2.12.1 and (1.6.3,IV), $|\partial_{\underline{t}}^{\perp}(g'\tau)|$ has the same order of magnitude as does $|(Dh\cdot\partial_{\underline{t}})^{\perp}(\gamma'\tau)|$. A rather intricate calculation, based on integrating an inequality on second derivatives, establishes this order of magnitude as $O(\tau^{k})$ (Lemma 2.15.2); here the second-order property of γ as a geodesic is essential. Using the estimate for $t(\tau)$ described above, this shows that

$$\left|\frac{d}{d\tau}Z(\tau)\right| = O(\tau^{-1+k}).$$

It follows that whenever $0 < \tau_1 \le \tau_2$,

$$d^L(Z\tau_1, Z\tau_2) = \int_{\tau_1}^{\tau_2} O(\tau^{-1+k})d\tau = O(\tau_2{}^{k}),$$

(Lemma 2.16); this implies that $\lim Z(\tau)$ exists, which proves Proposition 2.9 (see 2.17.5).

Now let β_1 and β_2 be paths C^1 at P. They are called "equivalent" (see (2.20)) if

$$\lim_{\sigma\to 0^+} d^M(\beta_1\sigma, \beta_2\sigma)/\sigma = 0,$$

where d^M is the intrinsic metric on M (see (2.4)). (I avoid the word "tangent" because when M is contained in some Riemannian manifold N as in Chapter 7, "tangency" could also refer to a property of paths in N; for paths in M which are C^1 at P, equivalence is a stronger relation than tangency. See Example 2.27.) T_PM is the set of equivalence classes of paths C^1 at P, with the metric

$$d^T(\hat{\beta}_1, \hat{\beta}_2) = \limsup_{\sigma\to 0^+} d^M(\beta_1\sigma, \beta_2\sigma)/\sigma,$$

β_1 and β_2 being any representatives of $\hat{\beta}_1$ and $\hat{\beta}_2$ respectively. The main theorem of the chapter is that $T_P M$ is a metric cone, and that any chart h at P induces a conical homeomorphism $T_P h$: $c^\infty L \to T_P M$ (Theorem 2.26).

(2.1) Remark. Let $(T_P M)^o$ be given the C^∞ structure and Riemannian metric induced from $(c^\infty L)^o$ by $T_P h$; then $e(h)_P = h \circ (T_P h)^{-1}$ is an h-dependent chart at P. One might conjecture that $\exp_P$, if it exists, is that choice of chart e_P: $c^{\underline{t}}(S_P M) \to M$ at P and metric on $S_P M$ (the unit sphere in $T_P M$) which come closest to giving an isometry near P, as in the non-singular case. I do not think $\exp_P$ can be characterized in these terms without differentiabliity assumptions beyond those considered in this paper. Theorem 6.16 is a partial result in this direction.

In this chapter I shall assume (1.6.1), (1.6.2) and I--V of (1.6.3).

Intrinsic Metrics

(2.2) Differentiable paths. Let M and P satisfy (1.6.1). A continuous path β: $[\nu,\lambda] \to M$ is called differentiable if $\mathrm{im}(\beta) \subseteq M^o$ and β is differentiable, or if $\mathrm{im}(\beta) = P$, or if (2.2.1) -- (2.2.3) hold:

(2.2.1) $\beta^{-1}(P)$ is a single point, usually arranged to be 0;

(2.2.2) β: $([\nu,\lambda] - \{0\}) \to M^o$ is differentiable;

(2.2.3) $|\beta'(\sigma)|$ is bounded on $[\nu,\lambda] - \{0\}$.

β is called C^m (m a positive integer or ∞) if in (2.2.2) the restriction of β is C^m; note that the higher derivatives of β need not be bounded in norm on $[\nu,\lambda] - \{0\}$.

β is said to go from $\beta(\nu)$ to $\beta(\lambda)$; and if $\beta^{-1}(P) \in (\nu,\lambda)$, then β is said to go through P.

(2.2.4) Convention. When a path has unspecified domain, it will be assumed to be [0,1].

(2.2.5) Notation. For each $(Z,t) \in (c^\infty L)^o$, set $a_{Z,t}(\sigma) = (Z,\sigma t)$; thus a: $[0,\infty) \to c^\infty L$. When $t \leq \underline{t}$, so $h(Z,t)$ is defined, say $X = h(Z,t)$; then $h \circ a_{Z,t}$, defined on $[0,\underline{t}/t]$, will be denoted $\alpha_{Z,t}$ or α_X. Under I -- III of (1.6.3), α_X is a C^∞ path from P to X, for every $X \in M^o$.

(2.3) Definitions. A continuous path β is piecewise differentiable (or piecewise C^m, respectively) if there is a finite subdivision of $[\nu,\lambda]$ into subintervals on each of which β is differentiable (or C^m). The length and energy of a piecewise C^1 path β are defined to be, respectively,

$$(2.3.1)\quad L(\beta) = \int_{\nu}^{\lambda} |\beta'(\sigma)|d\sigma;$$

$$(2.3.2)\quad E(\beta) = \int_{\nu}^{\lambda} |\beta'(\sigma)|^2 d\sigma;$$

these quantities are both finite.

(2.4) The metric d^M. Under hypotheses I -- III of (1.6.3) there is a piecewise C^∞ path between any two points of M, as follows from (2.2.5). The metric d^M on M is defined by

$$(2.4.1)\quad d^M(X,Y) = \inf\{L(\beta) \text{ for } \beta \text{ piecewise } C^1 \text{ from X to Y}\}.$$

Similarly, if L is a compact, closed Riemannian (n-1)-manifold, then its metric d^L is defined in the same way, with the extra condition

$$d^L(Z_1,Z_2) = \infty \text{ if } Z_1 \text{ and } Z_2 \text{ are in different path components of L.}$$

If $M = c\underline{t}L$, then $d^{c\underline{t}L}$ is defined in terms of d^L by (1.16.6); and in fact $d^M = d^{c\underline{t}L}$.

Observe that if β is a C^1 path from P, then

$$\begin{aligned}(2.4.2)\quad d^M(P,\beta(\sigma)) &\le L(\beta \restriction [0,\sigma]) \\ &\le \sigma \sup_{(0,\sigma]} |\beta'(\sigma^*)|, && \text{by (2.3.1)} \\ &= O(\sigma;\ \text{fixed } \beta) \quad \text{(see (1.12.2)), by (2.2.3).}\end{aligned}$$

(2.5) Intrinsic metrics. Now let $\delta\colon [\nu,\lambda] \to M$ be a continuous path. Following Aleksandroff [1], the length of δ is defined as follows. Let

$$(2.5.1)\quad \Pi\colon\ \nu = \sigma_0 < \sigma_1 < \dots < \sigma_m = \lambda$$

be a partition of $[\nu,\lambda]$; and for each Π, set

(2.5.2) $\sum_{\Pi}(\delta) = \sum_{i=1}^{m} d^M(\delta(\sigma_{i-1}),\delta(\sigma_i)).$

Then the <u>length</u>

(2.5.3) $L^*(\delta) = \sup_{\Pi}\{\sum_{\Pi}(\delta)\}.$

Define

(2.5.4) $d^*(X,Y) = \inf\{L^*(\delta)$ for δ continuous from X to Y$\}$,

The metric d^M is <u>intrinsic</u> if, for all X and Y in M,

(2.5.5) $d^M(X,Y) = d^*(X,Y).$

It will be shown (Proposition 2.9.1 and Corollary 2.16.9) that under hypotheses I--V of (1.6.3), d^M is indeed intrinsic.

The following four properties of L^* will be useful. The first is elementary, and the others are proved by Aleksandroff [1, Ch. 2].

(2.5.6) $L^*(\delta) \geq d^M(\delta\nu,\delta\lambda)$

(take Π: $\nu < \lambda$ in (2.5.3)).

(2.5.7) Let δ_n: $[\nu,\lambda] \to M$ converge uniformly to δ as $n \to \infty$; then $L^*(\delta) \leq \liminf L^*(\delta_n)$.

(2.5.8) Any continuous δ can be reparametrized (by a self-homeomorphism of $[\nu,\lambda]$) proportionally to arc length.

This means that the new path δ^* satisfies

$$L^*(\delta^* \restriction [\nu,\sigma]) = ((\sigma-\nu)/(\lambda-\nu))L^*(\delta).$$

(2.5.9) It follows that arc length is additive: if $\delta \in [\nu,\lambda]$, then

$$L^*(\delta \restriction [\nu,\lambda]) = L^*(\delta \restriction [\nu,\sigma]) + L^*(\delta \restriction [\sigma,\lambda]).$$

The primary goal of the rest of this section is to prove that d^M is an intrinsic metric, under the assumptions that M is complete and that I -- III of (1.6.3) hold (see Proposition 2.9.1).

LEMMA 2.6. Let $\delta: [\nu,\lambda] \to M$ be a differentiable path. Let $\mathcal{D}: [\nu,\lambda] \to \mathbb{R}$ and $d: [\nu,\lambda] \to \mathbb{R}$ be C^1 functions such that for all σ,

$$d(\sigma) \leq |\delta'(\sigma)| \leq \mathcal{D}(\sigma).$$

Then

$$\int_\nu^\lambda d(\sigma)d\sigma \leq L^*(\delta) \leq \int_\nu^\lambda \mathcal{D}(\sigma)d\sigma.$$

<u>Proof</u>. I shall first prove a special case of the lemma:

(2.6.1) Lemma 2.6 holds when $\mathrm{im}(\delta) \subseteq M^o$.

Fix $\varepsilon > 0$ and $\sigma^* \in [\nu,\lambda]$. Let $e = e_{\sigma^*}$ denote the exponential map of M^o at $\delta(\sigma^*)$, restricted to some ball $\tilde{U} = \tilde{U}_{\sigma^*}$ about $\delta(\sigma^*)$ in $T_{\delta\sigma^*}M$ on which e is a diffeomorphism. Then $De(\delta\sigma^*)$ is the identity map of $T_{\delta\sigma^*}M$; so if $\tilde{U}$ is chosen small enough, then e and e^{-1} are ε-close to being isometries. That is, for every $\tilde{Y} \in \tilde{U}$ and $v \in T_{\tilde{Y}}(T_{\delta\sigma^*}M)$

(2.6.2) $\quad ||De(Y)\cdot v| - |v|| \leq \varepsilon \min(|v|,|De(\tilde{Y})\cdot v|)$.

Let r_{σ^*} be chosen so small that

(2.6.3) $\quad B_{\sigma^*} = B(\delta\sigma^*,M;r_{\sigma^*})$ is contained in $e_{\sigma^*}(\tilde{U}_{\sigma^*})$ and satisfies (1.15.8).

Let $Y',Y'' \in B_{\sigma^*}$, and set $\tilde{Y}' = e^{-1}(Y')$, $\tilde{Y}'' = e^{-1}(Y'')$. Let $\tilde{\alpha}$ be the straight-line path from $\tilde{Y}'$ to $\tilde{Y}''$ in $\tilde{U}$,

$$\tilde{\alpha}(\tau) = (1-\tau)\tilde{Y}' + \tau\tilde{Y}''.$$

Then

(2.6.4) $\quad d^M(Y',Y'') \leq L(e\circ\tilde{\alpha})$, by (2.4.1)

$$= \int_0^1 |De(\tilde{\alpha}\tau)\cdot\tilde{\alpha}'\tau|d\tau$$

$$\leq (1+\varepsilon)\int_0^1 |\tilde{\alpha}'\tau|d\tau, \quad \text{by (2.6.2)}$$

$$= (1+\varepsilon)L(\tilde{\alpha})$$

$$= (1 + \varepsilon)|\tilde{Y}'' - \tilde{Y}'|.$$

On the other hand, by (1.15.8) there is a unique shortest weak geodesic (see (1.15.7)) β from Y' to Y'', and $\text{im}(\beta) \subseteq B_{\sigma*}$. Then $L(\beta) = d^M(Y',Y'')$, and

$$(2.6.5) \quad |\tilde{Y}'' - \tilde{Y}'| \leq L(e^{-1} \circ \beta)$$
$$= \int_0^1 |De^{-1}(\beta\tau) \cdot \beta'\tau| d\tau$$
$$\leq (1 + \varepsilon) \int_0^1 |\beta'\tau| d\tau, \qquad \text{by (2.6.2)}$$
$$= (1 + \varepsilon) L(\beta)$$
$$= (1 + \varepsilon) d^M(Y',Y'').$$

Combining (2.6.4) and (2.6.5) gives

$$(2.6.6) \quad (1 + \varepsilon)^{-1}|\tilde{Y}'' - \tilde{Y}'| \leq d^M(Y',Y'') \leq (1 + \varepsilon)|\tilde{Y}'' - \tilde{Y}'|.$$

Now let σ' and σ'' be such that $\text{im}(\delta \restriction [\sigma',\sigma'']) \subseteq B_{\sigma*}$; and set $Y' = \delta(\sigma')$, $Y'' = \delta(\sigma'')$. Then for $\sigma \in [\sigma',\sigma'']$, by (2.6.2),

$$(1 - \varepsilon) d(\sigma) \leq |(e^{-1} \circ \delta)'(\sigma)| \leq (1 + \varepsilon) \mathcal{D}(\sigma).$$

The mean-value theorem (see Dieudonne, [4, Ch. 8, §5]) applied to $e^{-1} \circ \delta$ on $[\sigma',\sigma'']$ gives

$$\int_{\sigma'}^{\sigma''} (1 - \varepsilon) d(\sigma) d\sigma \leq |\tilde{Y}'' - \tilde{Y}'| \leq \int_{\sigma'}^{\sigma''} (1 + \varepsilon) \mathcal{D}(\sigma) d\sigma.$$

By (2.6.6),

$$(2.6.7) \quad ((1 - \varepsilon)/(1 + \varepsilon)) \int_{\sigma'}^{\sigma''} d(\sigma) d\sigma \leq d^M(\delta\sigma',\delta\sigma'')$$
$$\leq (1 + \varepsilon)^2 \int_{\sigma'}^{\sigma''} \mathcal{D}(\sigma) d\sigma.$$

One can choose finitely many balls B_σ^* as in (2.6.3) whose union covers $\text{im}(\delta)$. Now any partition Π^* of $[\nu,\lambda]$ has a refinement Π as in

(2.5.1) such that for each $i = 1,\ldots,m$, $\mathrm{im}(\delta\restriction[\sigma_{i-1},\sigma_i])$ is contained in some $B_{\sigma*}$. From (2.6.7) and (2.5.2),

$$(2.6.8)\quad ((1-\varepsilon)/(1+\varepsilon))\int_{\nu}^{\lambda} d(\sigma)d\sigma \le \sum_{\Pi}(\delta) \le (1+\varepsilon)^2\int_{\nu}^{\lambda} \mathcal{D}(\sigma)d\sigma.$$

The system of suitable partitions Π is cofinal in the system of all partitions Π^* of $[\nu,\lambda]$. Hence, taking the limit in (2.6.8),

$$((1-\varepsilon)/(1+\varepsilon))\int_{\nu}^{\lambda} d(\sigma)d\sigma \le L^*(\delta) \le (1+\varepsilon)^2\int_{\nu}^{\lambda} \mathcal{D}(\sigma)d\sigma;$$

and since ε was arbitrary, this proves (2.6.1).

To prove Lemma 2.6 in general it is sufficient to consider the case that $\delta\colon\ [0,\lambda]\to M$ is a differentiable path from P. Let ν_n be a decreasing sequence, converging to 0, with $\nu_0 = \lambda$. Then

$$\sum_{i=1}^{\infty} L^*(\delta\restriction[\nu_{i-1},\nu_i]) \le \sum_{i=1}^{\infty}\int_{\nu_{i-1}}^{\nu_i} \mathcal{D}(\sigma)d\sigma, \text{ by (2.6.1)}$$

$$= \int_{0}^{\lambda} \mathcal{D}(\sigma)d\sigma.$$

So by the integral test,

$$\lim_{n\to\infty} L^*(\delta\restriction[\nu_n,\lambda]) = \sum_{i=1}^{\infty} L^*(\delta\restriction[\nu_{i-1},\nu_i])$$

exists; call this limit $L^{\#}(\delta)$. Then

$$\int_{\nu_n}^{\lambda} d(\sigma)d\sigma \le L^*(\delta\restriction[\nu_n,\lambda]) \le \int_{\nu_n}^{\lambda} \mathcal{D}(\sigma)d\sigma,$$

by (2.6.1), so in the limit,

$$\int_0^\lambda d(\sigma)d\sigma \leq L^{\#}(\delta) \leq \int_0^\lambda \mathcal{D}(\sigma)d\sigma.$$

Thus it remains to show that

$$(2.6.9) \quad L^{\#}(\delta) = L^*(\delta).$$

Now

$$\begin{aligned} L^*(\delta) &= L^*(\delta \restriction [\nu_n,\lambda]) + L^*(\delta \restriction [0,\nu_n]) \quad \text{by (2.5.9)} \\ &\geq L^*(\delta \restriction [\nu_n,\lambda]); \end{aligned}$$

so

$$(2.6.10) \quad L^*(\delta) \geq L^{\#}(\delta).$$

On the other hand, given any partition

$$\Pi: \quad 0 = \sigma_0 < \sigma_1 < \ldots < \sigma_m = \lambda,$$

choose n* so that $\sigma_0 < \nu_n < \sigma_1$ whenever $n \geq n^*$. Set

$$\Pi': \quad 0 = \sigma_0 < \nu_n < \sigma_1 < \ldots < \sigma_m = \lambda,$$

and

$$\Pi^*: \quad \nu_n < \sigma_1 < \ldots < \sigma_m = \lambda.$$

Then

$$\begin{aligned} \sum_{\Pi}(\delta) &\leq \sum_{\Pi'}(\delta) \\ &= d^M(P,\delta\nu_n) + \sum_{\Pi^*}(\delta \restriction [\nu_n,\lambda]) \\ &\leq d^M(P,\delta\nu_n) + L^*(\delta \restriction [\nu_n,\lambda]) \\ &\leq d^M(P,\delta\nu_n) + L^{\#}(\delta). \end{aligned}$$

Let $n \to \infty$; then, since δ is continuous, $d^M(P,\delta\nu_n) \to 0$. Therefore

$$\sum_{\Pi}(\delta) \leq L^{\#}(\delta)$$

and it follows that

$$L^*(\delta) \leq L^{\#}(\delta).$$

Together with (2.6.10) this proves (2.6.9), and with it, Lemma 2.6.

q.e.d.

COROLLARY 2.6.11. Let β be a piecewise C^1 path in M. Then $L^*(\beta) = L(\beta)$.

(2.7) Geodesics. Among piecewise C^1 paths β^*: $[\nu,\lambda] \to M$ from X to Y suppose there exists one, β, which minimizes energy; then β is called a geodesic. An equivalent definition is that $|\beta'(\sigma)|$, when defined, has a constant value, denoted $|\beta'|$, and that $L(\beta) = d^M(X,Y)$. For example if γ is a geodesic from P, then (2.4.2) can be strengthened to

(2.7.1) $d^M(P,\gamma\sigma) = \sigma|\gamma'|$.

(2.7.2) Caution. The reader is reminded of (1.15.7): that what is called a "geodesic" in traditional differential geometry will be called a "weak geodesic" in this work.

LEMMA 2.8. Let M be complete, and assume that I -- III of (1.6.3) hold. Then:

(1) For every X and Y in M there exists a shortest continuous path from X to Y.

(2) Let δ be a shortest continuous path parametrized by arc length; then that portion of δ which lies in M^o is a weak geodesic.

Proof of (1). Let δ_n be a sequence of continuous paths from X to Y with domain $[\nu,\lambda]$, each parametrized by arc length, as in (2.5.8), such that $L^*(\delta_n)$ converges to $d^*(X,Y)$ (see (2.5.4)). Then whenever $[\sigma',\sigma''] \subseteq [\nu,\lambda]$,

$$d^M(\delta_n\sigma',\delta_n\sigma'') \leq L^*(\delta_n \restriction [\sigma',\sigma'']), \qquad \text{by (2.5.6)}$$

$$\leq ((\sigma'' - \sigma')/(\lambda - \nu))L^*(\delta_n),$$

by (2.5.8). Hence the family $\{\delta_n\}$ is equicontinuous. Since M is complete, there exists a subsequence $\{\delta_{n*}\}$ which converges uniformly to a continuous path δ from X to Y. Then

$$L^*(\delta) \le \liminf L^*(\delta_{n*}), \quad \text{by (2.5.7)}$$
$$= d^*(X,Y);$$

so by (2.5.4), $L^*(\delta) = d^*(X,Y)$, which proves (1).

<u>Proof of (2)</u>. Let δ be as in the assertion and say $\delta(\sigma^*) \in M^o$. Let B be a ball about $\delta(\sigma^*)$ in M^o satisfying (1.15.8), and let $[\sigma',\sigma'']$ be a neighborhood of σ^* in $[\nu,\lambda]$ such that $\mathrm{im}(\delta \restriction [\sigma',\sigma'']) \subseteq B$. Let β be the (unique) shortest weak geodesic from $\delta(\sigma')$ to $\delta(\sigma'')$ in M^o with domain $[\sigma',\sigma'']$. Then $\mathrm{im}(\beta) \subseteq B$; and whenever $[\underline{\sigma}',\underline{\sigma}''] \subseteq [\sigma',\sigma'']$,

(2.8.3) $L(\beta \restriction [\underline{\sigma}',\underline{\sigma}'']) = d^M(\beta\underline{\sigma}',\beta\underline{\sigma}'')$.

Suppose now that for some $\sigma \in (\sigma',\sigma'')$, $\delta(\sigma) \notin \mathrm{im}(\beta)$. Then, since $\delta(\sigma) \in B$, there are unique shortest weak geodesics β' and β'' from $\delta(\sigma')$ to $\delta(\sigma)$ and from $\delta(\sigma)$ to $\delta(\sigma'')$ with domains $[\sigma',\sigma]$ and $[\sigma,\sigma'']$ respectively. The combined piecewise C^1 path $\beta' + \beta''$: $[\sigma',\sigma''] \to M^o$ must be strictly longer than β, by (1.15.8). Therefore, applying (2.8.3) to β, β' and β'',

(2.8.4) $d^M(\delta\sigma',\delta\sigma) + d^M(\delta\sigma,\delta\sigma'') \underset{\neq}{>} d^M(\delta\sigma',\delta\sigma'')$.

But now

$$\begin{aligned} L^*(\delta \restriction [\sigma',\sigma'']) &\ge d^M(\delta\sigma',\delta\sigma) + d^M(\delta\sigma,\delta\sigma''), && \text{by (2.5.2) and (2.5.3)} \\ &\underset{\neq}{>} d^M(\delta\sigma',\delta\sigma''), && \text{by (2.8.4)} \\ &= L(\beta), && \text{by (2.8.3)} \\ &= L^*(\beta), \end{aligned}$$

by Corollary 2.6.11. Hence δ can be shortened (in view of (2.5.9)) by replacing $\delta \restriction [\sigma',\sigma'']$ with β, contradicting the definition of δ. Therefore $\mathrm{im}(\delta) \subseteq \mathrm{im}(\beta)$. Moreover

(2.8.5) $L^*(\delta \restriction [\sigma',\sigma'']) = L(\beta)$.

Now suppose that for some $\sigma \in (\sigma',\sigma")$, $\delta(\sigma) \neq \beta(\sigma)$. Set $\underline{\sigma} = \beta^{-1}(\delta\sigma)$ and say $\underline{\delta} \underset{\neq}{>} \sigma$; the argument in the contrary case will be similar. Then

$$\begin{aligned} L^*(\delta\restriction[\sigma',\sigma"]) &\geq d^M(\delta\sigma',\delta\sigma"), && \text{by (2.5.6)} \\ &= d^M(\beta\sigma',\beta\underline{\sigma}) \\ &= L(\beta\restriction[\sigma',\underline{\sigma}]) \\ &\underset{\neq}{>} L(\beta\restriction[\sigma',\sigma]) \\ &= ((\sigma-\sigma')/(\sigma"-\sigma'))L(\beta) \\ &= ((\sigma-\sigma')/(\sigma"-\sigma'))L^*(\delta\restriction[\sigma',\sigma"]), \end{aligned}$$

by (2.8.5). But this contradicts the fact that δ is parametrized proportionally to arc length (see (2.5.8)). Therefore $\delta\restriction[\sigma',\sigma"] = \beta$; that is, δ is a weak geodesic near $\delta(\sigma^*)$. Assertion (2) now follows.

q.e.d.

PROPOSITION 2.9. Assume that M is complete and that I -- III of (1.6.3) hold. Then:

(1) The metric d^M is intrinsic in the sense of (2.5.5);

(2) For every $X,Y \in M$ there exists a geodesic from X to Y;

(3) That portion of any geodesic which lies in M^O is a weak geodesic.

<u>Proof of (1)</u>. It follows from the definitions of $d^M(X,Y)$ and $d^*(X,Y)$ (see (2.4.1) and (2.5.4)) and from Corollary 2.6.11 that for all $X,Y \in M$

(2.9.4) $d^M(X,Y) \geq d^*(X,Y)$.

Conversely, by Lemma 2.8.1 and (2.5.8), there exists a continuous path δ from X to Y with domain $[\nu,\lambda]$, parametrized proportionally to arc length, such that

(2.9.5) $d^*(X,Y) = L^*(\delta)$.

Say $P = \delta^{-1}(0)$; if $0 \notin [\nu,\lambda]$, this means $\mathrm{im}(\delta) \subseteq M^O$. Then on each component of $[\nu,\lambda] - \{0\}$, δ is a weak geodesic by Lemma 2.8.2. So on each such component, δ is C^∞ and $|\delta'(\sigma)|$ is constant. It follows that δ is piecewise C^∞. Therefore $L(\delta)$ is defined, and

$$(d^M(X,Y) \leq L(\delta)$$

$$= L^*(\delta), \quad \text{by Corollary 2.6.11}$$
$$= d^*(X,Y),$$

by (2.9.5). Together with (2.9.4), this proves (1).

<u>Proof of (2)</u>. The argument above shows that any δ satisfying (2.9.5) is in fact a geodesic from X to Y.

<u>Proof of (3)</u>. Let β be a geodesic from X to Y; then

$$\begin{aligned} L^*(\beta) &= L(\beta), \quad \text{by Corollary 2.6.11} \\ &= d^M(X,Y) \\ &= d^*(X,Y), \end{aligned}$$

by (1). So β is a shortest continuous path from X to Y; and since β is parametrized proportionally to arc length, (3) follows from Lemma 2.8.2.

q.e.d.

LEMMA 2.10. Let $\beta_n\colon [\nu,\lambda] \to M$ be a sequence of geodesics which converge uniformly (pointwise). Assume M is complete and satisfies I -- III of (1.6.3). Then $\lim_{n\to\infty} \beta_n$ is a geodesic.

<u>Proof</u>. Set $\beta = \lim_{n\to\infty} \beta_n$ (which exists since M is complete). Let $\sigma \in [\nu,\lambda]$. Then each $\beta_n \upharpoonright [\nu,\sigma]$ is a geodesic and hence, by the proof of Proposition 2.9.3, is a shortest continuous path from $\beta_n(\nu)$ to $\beta_n(\sigma)$. Now

$$\begin{aligned} (2.10.1) \quad L^*(\beta \upharpoonright [\nu,\sigma]) &\leq \liminf L^*(\beta_n \upharpoonright [\nu,\sigma]), \quad \text{by (2.5.8)} \\ &= \liminf d^*(\beta_n\nu, \beta_n\sigma) \\ &= d^*(\beta\nu, \beta\sigma), \end{aligned}$$

since $d^* = d^M$ is continuous in its two parameters. By definition of d^*, (2.5.4), it follows that the inequality in (2.10.1) is an equality. In particular, setting $\sigma = \lambda$ gives

$$(2.10.2) \quad L^*(\beta) = d^*(\beta\nu, \beta\lambda).$$

Now

$$L^*(\beta\restriction[\nu,\sigma]) = \liminf L^*(\beta_n\restriction[\nu,\sigma])$$

$$= \liminf ((\sigma - \nu)/(\lambda - \nu))L^*(\beta_n),$$

since β_n is parametrized proportionally to arc length,

$$= ((\sigma - \nu)/(\lambda - \nu)) \liminf d^*(\beta_n\nu,\beta_n\lambda)$$

$$= ((\sigma - \nu)/(\lambda - \nu))d^*(\beta\nu,\beta\lambda)$$

$$= ((\sigma - \nu)/(\lambda - \nu))L^*(\beta),$$

by (2.10.2). Thus β is a shortest continuous path parametrized proportionally to arc length. The proof of Proposition 2.9.2 now shows that β is a geodesic. q.e.d.

Paths C^1 at P

(2.11) Definition. Let L satisfy (1.6.2). A path $b(\sigma) = (Z(\sigma),t(\sigma))$ from c in cL is called C^1 at c if it is C^1 in the sense of (2.2) and

$$\lim_{\sigma\to 0^+} (Z(\sigma),t(\sigma)/\sigma) \quad \text{exists in} \quad c^\infty L. \quad \text{(See Figure 1.)}$$

Let h: $cL \to M$ be a chart (compare (1.16.10)); then a path β from P is C^1 at P with respect to h if, on an initial segment of β which lies in

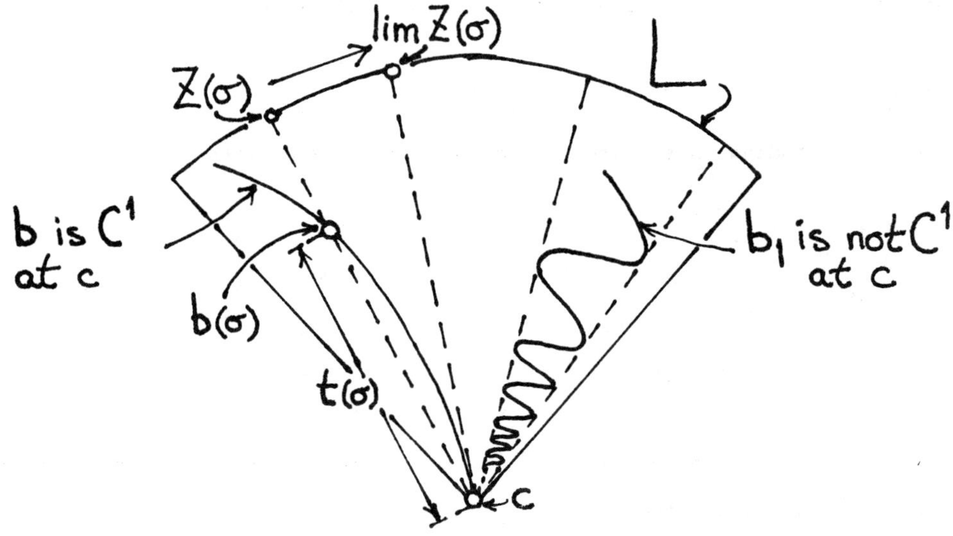

Figure 1

$\mathrm{im}(h)$, $h^{-1}\circ\beta$ is C^1 at c.

The main result of this section is that the last definition is independent of the choice of h (Proposition 2.14). Once this result is proved, β will be called simply $\underline{C^1 \text{ at } P}$.

LEMMA 2.12. If β is C^1 at P with respect to h, then β is C^1 in the sense of (2.2). (Proved in (2.16.6).)

PROPOSITION 2.13. Let γ be a geodesic from P and let h be a chart at P. Then γ is C^∞ and is C^1 at P with respect to h. (Proved in (2.22).)

PROPOSITION 2.14. Let h and h* be charts at P and let β be a path from P. Then β is C^1 at P with respect to h if and only if it is so with respect to h*. (Proved in (2.24).)

(2.15) Notation. Given a chart h of M at P, define a vector field $\underline{\rho}$ on $M^o = h((cL)^o)$ by

(2.15.1) $\underline{\rho}(X) = Dh(h^{-1}X)\cdot\underline{\partial}_t$

(compare (1.14.4) and (1.16.5)). The next lemma shows there exists $\underline{t} > 0$ such that $\underline{\rho}$ is nowhere 0 on $h(c^{\underline{t}}L)^o)$. Changing h as in (1.16.10) gives a new chart h* such that $h^*(cL) = h(c^{\underline{t}}L)$; and now $\underline{\rho}^*$ is nowhere 0 on $h^*((cL)^o)$. Thus I may assume (after Lemma 2.16.2) that

(2.15.2) $\underline{\rho}$ is nowhere 0 on $h((cL)^o)$.

The corresponding unit vector field on M^o will be denoted

(2.15.3) $\underline{\rho}^* = |\underline{\rho}|^{-1}\underline{\rho}$.

Also set

(2.15.4) $\tau(X) = d^M(P,X)$.

LEMMA 2.16. (1) There exist $H \geq h > 0$ such that whenever $(Z^*,t^*) \in (cL)^o$ and $u^* \in T_{(Z^*,t^*)}cL$, then

$$h|u^*| \leq |Dh\cdot u^*| \leq H|u^*|.$$

(2) $|\underline{\rho}|(h(Z^*,t^*)) = 1 + O(t^{*2k})$,

where $k > 0$ is as in (1.6.3,V).

Proof of (1). In view of (1.6.3,IV) it is enough to prove (1) when $\langle u^*,\underline{\partial}_t\rangle = 0$ and to prove (2). So assume $\langle u^*,\underline{\partial}_t\rangle = 0$ and $u^* \neq 0$. Set $a = a_{Z^*,1}$ (see (2.2.5)), and let $\underline{u}$ be the vector field along a defined by

$$\underline{u}(t) = Di_t \circ (Di_{t^*})^{-1} \cdot u^*$$

(see (1.16.2)); then $\underline{u}(t^*) = u^*$ and $|\underline{u}(t)| = t|\underline{u}(1)|$. Set $\alpha = h \circ a$ and $\underline{\psi} = Dh \cdot \underline{u}$, a vector field along α. From $a'(t) = \underline{\partial}_t(Z^*,t)$ follows that $\alpha'(t) = \underline{\rho}(h(Z^*,t))$.

Now $\underline{u}$ is nowhere 0, $\langle \underline{u},a'\rangle = \langle \underline{u},\underline{\partial}_t\rangle \equiv 0$ and $[\underline{u},\underline{\partial}_t] \equiv 0$. Hence $\underline{\psi}$ is nowhere 0, $\langle \underline{\psi},\underline{\rho}\rangle = \langle \underline{\psi},\alpha'\rangle \equiv 0$ (by (1.6.3,IV)) and $[\underline{\psi},\underline{\rho}] \equiv 0$. The last identity implies (see Hicks [8,§6.2])

$$D^M_{\underline{\rho}(\alpha t)}\underline{\psi} = D^M_{\underline{\psi}(t)}\underline{\rho} \quad \text{(compare (1.15.3)).}$$

Now

$$\begin{aligned}\frac{d}{dt}|\underline{\psi}(t)| &= \partial_{\underline{\rho}(\alpha t)}|\underline{\psi}|, \qquad \text{see (1.14.4)}\\ &= |\underline{\psi}(t)|^{-1}\langle \underline{\psi}(t), D^M_{\underline{\rho}(\alpha t)}\underline{\psi}\rangle\\ &= |\underline{\psi}(t)|^{-1}\langle \underline{\psi}(t), D^M_{\underline{\psi}(t)}\underline{\rho}\rangle\\ &= |\underline{\psi}(t)|(t^{-1} + O(t^{-1+2k})),\end{aligned}$$

by (1.6.3,V). Thus there exists $A > 0$, independent of (Z^*,t^*) and of u^*, such that

$$(t^{-1} - At^{-1+2k})|\underline{\psi}(t)| \leq \frac{d}{dt}|\underline{\psi}(t)| \leq (t^{-1} + At^{-1+2k})|\underline{\psi}(t)|.$$

The o.d.e. on $(0,1]$

$$y' = (t^{-1} + At^{-1+2k})y, \quad y(1) = |\underline{\psi}(1)|$$

has as solution

$$y = t|\psi(1)|\exp(-A(1 - t^{2k})/2k).$$

Hence (see Hartman [7, Ch. 3,§4])

(2.16.3) $|\underline{\psi}(t)| \geq y(t) \geq t|\underline{\psi}(1)|\exp(-A/2k)$.

Since L is compact, there exist $H^L \geq h^L > 0$ such that whenever $Z \in L$ and $u \in T_Z L$, then

$$h^L|u| \leq |Dh\cdot u| \leq H^L|u|.$$

Substitute $Z = Z^*$, $u = \underline{u}(1)$ and $Dh\cdot u = \underline{\psi}(1)$; then (2.16.3) gives

$$\begin{aligned} |Dh\cdot u^*| &= |\underline{\psi}(t^*)| \\ &\geq t^* h^L|\underline{u}(1)|\exp(-A/2k) \\ &= h^L \exp(-A/2k)|u^*| \\ &= h|u^*|, \end{aligned}$$

say, thus defining h. A similar argument shows that

$$\begin{aligned} |Dh\cdot u^*| &\leq H^L \exp(A/2k)|u^*| \\ &= H|u^*|, \end{aligned}$$

say, thus defining H. This proves (1) when $\langle u^*, \underline{\partial}_t\rangle = 0$.

Proof of (2). Let α be as in the proof of (1), with $\alpha'(t) = \underline{\rho}(\alpha t)$. Then

(2.16.4)
$$\begin{aligned} \frac{d}{dt}|\underline{\rho}(\alpha t)| &= \partial_{\underline{\rho}(\alpha t)}|\underline{\rho}| \\ &\leq |D^M_{\underline{\rho}(\alpha t)}\underline{\rho}| \\ &= |\underline{\rho}(\alpha t)|\,O(t^{-1+2k}), \end{aligned}$$

by (1.6.3,V). Hence there exists $A > 0$ such that

$$-At^{-1+2k}|\underline{\rho}(\alpha t)| \leq \frac{d}{dt}|\underline{\rho}(\alpha t)| \leq At^{-1+2k}|\underline{\rho}(\alpha t)|.$$

By (1.6.3,III), $\lim_{t\to 0^+} |\underline{\rho}(\alpha t)| = 1$.

The o.d.e. on (0,1]

$$y' = At^{-1+2k}y, \quad \lim_{t\to 0^+} y(t) = 1$$

has as solution

$$y = \exp(At^{2k}/2k).$$

As in the proof of (1) it follows that

$$\exp(-At^{2k}/2k) \leq |\underline{\rho}(\alpha t)| \leq \exp(At^{2k}/2k).$$

Since the left and right hand terms are both $1 + 0(t^{2k})$, assertion (2) follows. q.e.d.

COROLLARY 2.16.5. Let h: $cL \to M$ be a chart at P and let β be a path from P in $im(h)$. Then β is C^1 (or piecewise C^1) if and only if $h^{-1}o\beta$ is C^1 (or piecewise C^1, respectively).

Proof. The only point requiring examination is that in the C^1 case (2.2.3) holds for β if and only if it holds for $h^{-1}o\beta$; and this follows from Lemma 2.16.1. q.e.d.

(2.16.6) Proof of Lemma 2.12. This is a special case of Corollary 2.16.5. q.e.d.

COROLLARY 2.16.7. Let h: $cL \to M$ be a chart at P, and let H and h be as in Lemma 2.16.1. Then:

(1) Given a piecewise C^1 path b in cL, then

$$hL(b) \leq L(hob) \leq HL(b);$$

$$h^2E(b) \leq E(hob) \leq H^2E(b)$$

(2) Given a piecewise C^1 path β in $im(h)$, then

$$H^{-1}L(\beta) \leq L(h^{-1}o\beta) \leq h^{-1}L(\beta);$$

$$H^{-2}E(\beta) \leq E(h^{-1}o\beta) \leq h^{-2}E(\beta).$$

COROLLARY 2.16.8. Let h: $cL \to M$ be a chart at P. Let $X,Y \in im(h)$ be such that $\tau(X),\tau(Y) < (1/2)d^M(P,M - \{im(h)\})$. Then:

(1) $hd^{cL}(h^{-1}X,h^{-1}Y) \leq d^M(X,Y) \leq Hd^{cL}(h^{-1}X,h^{-1}Y)$;

(2) $H^{-1}d^M(X,Y) \leq d^{cL}(h^{-1}X,h^{-1}Y) \leq h^{-1}d^M(X,Y)$.

Proof. The inequalities on $\tau(X)$ and $\tau(Y)$ (see (2.15.4)) imply that in the definition (2.4.1) of $d^M(X,Y)$ one need only look at piecewise C^1 paths from X to Y that lie in im(h). The result now follows from Corollary 2.16.7. q.e.d.

COROLLARY 2.16.9. M is complete in the metric d^M.

Proof. This follows from Corollary 2.16.8 and the fact that cL is complete in the metric d^{cL}.

PROPOSITION 2.17. Assume that I -- V of (1.6.3) hold. Then:

(1) The metric d^M is intrinsic;

(2) For every $X,Y \in M$ there exists a geodesic from X to Y;

(3) That portion of any geodesic which lies in M^o is a weak geodesic.

Proof. This follows from Corollary 2.16.9 and Proposition 2.9. q.e.d.

COROLLARY 2.17.4. For every $X \in M$ there exists a geodesic γ_X from P to X with domain [0,1]. Thus property (1.1) of an exponential map holds for M at P.

LEMMA 2.18. Let γ be a geodesic from P in M parametrized by arc length on the domain $[0,\tau_1]$. Set $g = h^{-1}o\gamma$, and for $\tau > 0$ write $g(\tau) = (Z(\tau),t(\tau))$. Then:

(1) $t(\tau) = \tau(1 + O(\tau^{2k}))$;

(2) $\tau = t(\tau)(1 + O(t(\tau)^{2k}))$.

Proof. Fix $\tau^* > 0$ and set $t^* = t(\tau^*)$. Let

$$t^{\#} = \max\{t(\tau) \text{ for } 0 < \tau \leq \tau^*\}.$$

Since $\lim_{\tau\to 0^+} t(\tau) = 0$, there exists $\tau^{\#} \in (0,\tau^*]$ such that $t(\tau^{\#}) = t^{\#}$. Set $a = a_{Z(\tau^*),1} \restriction [0,\tau^*]$ (see (2.2.5)) and $\alpha = hoa$; so α is an integral curve of $\underline{\rho}$. (See Figure 2.) Now α and $\gamma \restriction [0,\tau^*]$ have the same endpoints. Since γ is a geodesic parametrized by arc length in M,

$$\begin{aligned}\tau^* &= L(\gamma \restriction [0,\tau^*]) \\ &\leq L(\alpha) \\ &= \int_0^{t^*} |\alpha'(t)|\,dt \\ &= \int_0^{t^*} |\underline{\rho}(\alpha t)|\,dt \\ &= \int_0^{t^*} (1 + O(t^{2k}))dt, \quad \text{by Lemma 2.16.2} \\ &= t^*(1 + O(t^{\#2k})).\end{aligned}$$

Hence

(2.18.3) $t^* \geq \tau^*(1 + O(t^{\#2k}))$.

Now

(2.18.4) $\frac{d}{d\tau}t(\tau) = \langle g'(\tau), \underline{\partial}_t\rangle$.

By (1.6.3,IV) (compare (1.13.5)),

$$Dh\cdot(\underline{\partial}_t^{\parallel}(g'\tau)) = \underline{\rho}^{\parallel}(\gamma'\tau);$$

that is,

$$\begin{aligned}Dh(\langle g'\tau, \underline{\partial}_t\rangle \underline{\partial}_t) &= \langle \gamma'\tau, \underline{\rho}^*\rangle \underline{\rho}^*(\gamma\tau) \quad \text{(see (2.15.3))} \\ &= (\langle \gamma'\tau, \underline{\rho}\rangle / |\underline{\rho}(\gamma\tau)|^2)\underline{\rho}(\gamma\tau).\end{aligned}$$

Since $Dh\cdot\underline{\partial}_t = \underline{\rho}$, it follows that

$$\langle g'\tau, \underline{\partial}_t\rangle = \langle \gamma'\tau, \underline{\rho}\rangle / |\underline{\rho}(\gamma\tau)|^2.$$

Hence, using Lemma 2.16.2 in (2.18.4),

(2.18.5) $\frac{d}{d\tau}t(\tau) = \langle \gamma'\tau, \underline{\rho}\rangle(1 + O(t(\tau)^{2k}))$.

Thus, since $|\gamma'| = 1$,

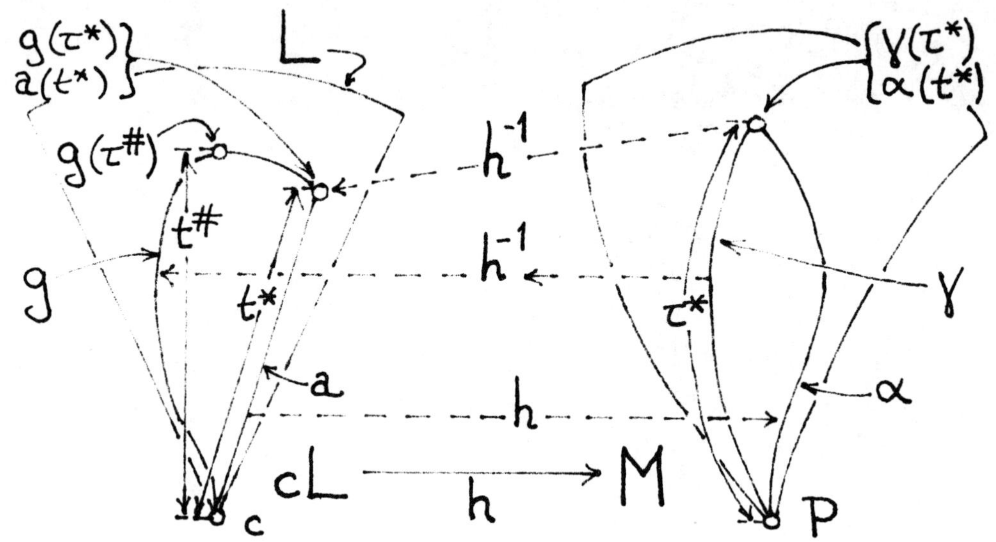

Figure 2

$$\left|\frac{d}{d\tau}t(\tau)\right| \leq 1 + 0(t^{\#2k}).$$

Integrating from $\tau = 0$ to $\tau = \tau^*$ gives

$$t^* \leq \tau^*(1 + 0(t^{\#2k})),$$

which, in combination with (2.18.3) shows that

(2.18.6) $t^* = \tau^*(1 + 0(t^{\#2k}))$.

In particular, set $\tau^* = \tau^\#$; then $t^* = t^\#$, so

$$t^\# = \tau^\#(1 + 0(t^{\#2k})).$$

Substituting this into (2.18.6) gives

$$\begin{aligned} t^* &= \tau^*(1 + 0(\tau^{\#2k})) \\ &= \tau^*(1 + 0(\tau^{*2k})), \end{aligned}$$

since $\tau^\# \leq \tau^*$. This proves (1); and (2) follows immediately. q.e.d.

COROLLARY 2.18.7. Let $T > 0$ be given and let γ be a geodesic from P in M with $|\gamma'| \leq T$. For $\sigma > 0$, write $h^{-1} \circ \gamma(\sigma) = (Z(\sigma), t(\sigma))$. Then

$$t(\sigma) = \sigma|\gamma'|(1 + O((\sigma T)^{2k})).$$

COROLLARY 2.18.8. Let $X \in M^o$; then $|\underline{\rho}(X)| = 1 + O(\tau(X)^{2k})$, where $\tau(X)$ is as in (2.15.4).

(2.19) Notation. Let γ be a geodesic from P and let $\underline{\rho}^*$ be the unit vector field of (2.15.3). For $\sigma > 0$, set

(2.19.1) $$\underline{\rho}(\gamma\sigma)^{\parallel}(\gamma'\sigma) = \phi(\sigma)\underline{\rho}^*(\gamma\sigma), \quad \text{and}$$

(2.19.2) $$\underline{\rho}(\gamma\sigma)^{\perp}(\gamma'\sigma) = \underline{\chi}(\sigma).$$

Thus ϕ is a scalar function, and $\underline{\chi}$ a vector field, along γ;

(2.19.3) $\phi(\sigma) = \langle \gamma'\sigma, \underline{\rho}^*(\gamma\sigma)\rangle$, and

(2.19.4) $\gamma'(\sigma) = \phi(\sigma)\underline{\rho}^*(\gamma\sigma) + \underline{\chi}(\sigma)$.

LEMMA 2.20. Let the notation be as in (2.19), and assume γ is parametrized by arc length τ. Then:

(1) $\phi(\tau) = 1 + O(\tau^{2k})$;

(2) $|\underline{\chi}(\tau)| = O(\tau^k)$.

Proof of (1). The first step is to derive the differential inequality (2.20.3) below. For $\tau > 0$,

$$\begin{aligned}\frac{d}{d\tau}\phi(\tau) &= D^M_{\gamma'\tau}\langle \gamma', \underline{\rho}^*\rangle \\ &= \langle \gamma'\tau, D^M_{\gamma'}\underline{\rho}^*\rangle,\end{aligned}$$

since γ is a weak geodesic in M^o in the sense of (1.15.7), by Proposition 2.9.3. Applying (2.19.4),

$$\frac{d}{d\tau}\phi(\tau) = \phi(\tau)\langle \underline{\rho}^*, D^M_{\gamma'\tau}\underline{\rho}^*\rangle + \langle \underline{\chi}, D^M_{\gamma'\tau}\underline{\rho}^*\rangle.$$

But $\underline{\rho}^*$ is a unit vector field, so $\langle \underline{\rho}^*, D^M_{\gamma'\tau}\underline{\rho}^*\rangle = 0$. Thus

$$\frac{d}{d\tau}\phi(\tau) = \langle \underline{\chi}, D^M_{\gamma'\tau}\underline{\rho}^*\rangle$$

$$= \phi(\tau)\langle \underline{\chi}, D^M_{\underline{\rho}*(\gamma\tau)}\underline{\rho}*\rangle + \langle \underline{\chi}, D^M_{\underline{\chi}(\tau)}\underline{\rho}*\rangle.$$

Now given any vector $\psi \in T_{\gamma\tau}M$,

$$D^M_{\psi}\underline{\rho}* = |\underline{\rho}(\gamma\tau)|^{-1}D^M_{\psi}\underline{\rho} + \partial_{\psi}(|\underline{\rho}|^{-1})\underline{\rho}(\gamma\tau).$$

The second term is orthogonal to $\underline{\chi}(\gamma\tau)$; so

$$\frac{d}{d\tau}\phi(\tau) = \phi(\tau)|\underline{\rho}(\gamma\tau)|^{-1}\langle \underline{\chi}, D^M_{\underline{\rho}*(\gamma\tau)}\underline{\rho}\rangle + |\underline{\rho}(\gamma\tau)|^{-1}\langle \underline{\chi}, D^M_{\underline{\chi}(\tau)}\underline{\rho}\rangle.$$

Therefore, since $|\phi(\tau)| \leq 1$,

$$\left|\frac{d}{d\tau}\phi(\tau)\right| \leq |\underline{\rho}(\gamma\tau)|^{-1}|\underline{\chi}(\tau)|(|D^M_{\underline{\rho}*(\gamma\tau)}\underline{\rho}| + |D^M_{\underline{\chi}(\tau)}\underline{\rho}|).$$

Using (1.6.3,IV), since $\underline{\rho}^{\perp}(\underline{\rho}*(\gamma\tau)) = 0$ and $|\underline{\rho}^{\perp}(\underline{\chi}(\tau))| \leq |\underline{\chi}(\tau)|$, it follows that

$$\left|\frac{d}{d\tau}\phi(\tau)\right| \leq |\underline{\rho}(\gamma\tau)|^{-1}|\underline{\chi}(\tau)|\{0(\tau^{-1+2k}) + |\underline{\chi}(\tau)|(\tau^{-1} + 0(\tau^{-1+2k}))\}$$

$$= |\underline{\chi}(\tau)|0(\tau^{-1+2k}) + |\underline{\chi}(\tau)|^2 0(\tau^{-1}),$$

by Corollary 2.18.8. In other words, there exists $A > 0$ such that

$$(2.20.3)\qquad \left|\frac{d}{d\tau}\phi(\tau)\right| \leq A\{(1 - \phi(\tau)^2)^{\frac{1}{2}}\tau^{-1+2k} + (1 - \phi(\tau)^2)\tau^{-1}\};$$

here I have substituted for $|\underline{\chi}(\tau)|$ from

$$(2.20.4)\qquad \begin{aligned} |\underline{\chi}(\tau)|^2 + \phi(\tau)^2 &= |\gamma'(\tau)|^2, && \text{by (2.19.1) and (2.19.2)} \\ &= 1, && \text{by hypothesis on } \gamma. \end{aligned}$$

To integrate (2.20.3) requires the preliminary estimate (2.20.5). Fix $\tau^* > 0$, and for $0 < \tau \leq \tau^*$ write $h^{-1}o\gamma(\tau) = (Z(\tau), t(\tau))$. Then

$$\frac{d}{d\tau}t(\tau) = \phi(\tau)|\underline{\rho}(\gamma\tau)|(1 + 0(t(\tau)^{2k})),$$

by (2.18.5), (2.19.3) and (2.15.3),

$$= \phi(\tau)(1 + 0(\tau^{*2k})),$$

by Lemma 2.18.1 and Corollary 2.18.8. Hence

$$t(\tau^*) = (1 + O(\tau^{*2k})) \int_0^{\tau^*} \phi(\tau)d\tau.$$

Since the left-hand side is, by Lemma 2.18.1 again, $\tau^*(1 + O(\tau^{*2k}))$, it follows that

$$\text{(2.20.5)} \qquad \int_0^{\tau_*} \phi(\tau)d\tau = \tau^*(1 + O(\tau^{*2k})).$$

We can now estimate the integral of the right-hand side of (2.20.3). Since $|\phi(\tau)| \le 1$, $1 - \phi(\tau)^2 \le 2(1 - \phi(\tau))$; so from (2.20.5),

$$\text{(2.20.6)} \qquad \int_0^{\tau_2} (1 - \phi(\tau)^2)d\tau = O(\tau_2^{1+2k}).$$

Integrating by parts shows that, for $0 < \tau_1 \le \tau_2$,

$$\text{(2.20.7)} \qquad \int_{\tau_1}^{\tau_2} (1 - \phi(\tau)^2)\tau^{-1}d\tau = O(\tau_2^{2k});$$

this estimates the integral of the first term on the right-hand side of (2.20.3). As for the other term, by the Schwartz inequality,

$$\left[\int_0^{\tau_2} (1 - \phi(\tau)^2)^{\frac{1}{2}}dt\right]^2 \le \int_0^{\tau_2} (1 - \phi(\tau)^2)d\tau \int_0^{\tau_2} 1\, d\tau$$
$$= O(\tau_2^{2+2k}),$$

by (2.20.6); hence

$$\int_0^{\tau_2} (1 - \phi(\tau)^2)^{\frac{1}{2}}d\tau = O(\tau_2^{1+k}).$$

Integrating by parts again gives the following estimate for the integral of the second term on the right-hand side of (2.20.3):

$$(2.20.8)\quad \int_{\tau_1}^{\tau_2} (1 - \phi(\tau)^2)^{\frac{1}{2}}\tau^{-1+k}d\tau = 0(\tau_2{}^{2k}).$$

Now integrate (2.20.3) from τ_1 to τ_2, applying (2.20.7) and (2.20.8). This yields

$$|\phi(\tau_2) - \phi(\tau_1)| \le \int_{\tau_1}^{\tau_2} |\tfrac{d}{d\tau}\phi(\tau)|d\tau = 0(\tau_2{}^{2k}).$$

Therefore $\phi(0) = \lim_{\tau\to 0^+} \phi(\tau)$ exists, and

$$(2.20.9)\quad \phi(\tau_2) = \phi(0) + 0(\tau_2{}^{2k}).$$

Hence

$$\int_0^{\tau^*} \phi(\tau_2)d\tau_2 = \tau^*(\phi(0) + 0(\tau^{*2k})).$$

Comparing this with (2.20.5) shows that $\phi(0) = 1$. Now assertion (1) follows from (2.20.9).

Proof of (2). This follows immediately from (1) and (2.20.4). q.e.d.

LEMMA 2.21. Let γ be a geodesic from P parametrized by arc length. Let h be a chart at P and assume $\mathrm{im}(\gamma) \subseteq \mathrm{im}(h)$. Set $g = h^{-1}\circ\gamma$, and for $\tau > 0$ write $g(\tau) = (Z(\tau),t(\tau))$. Let a_τ stand for $a_{g(\tau)}$ (see (2.2.5)). Then for all $\tau > 0$ and $\lambda \in (0,1]$:

(1) $d^L(Z(\lambda\tau),Z(\tau)) = 0(\tau^k)$;

(2) $d^{cL}(g(\lambda\tau),a_\tau(\lambda)) = \lambda 0(\tau^{1+k})$.

(See Figure 3.)

Proof of (1).

$$|\tfrac{d}{d\tau}Z(\tau)| = |\partial_t{}^{\perp}(g'\tau)|/t(\tau)$$

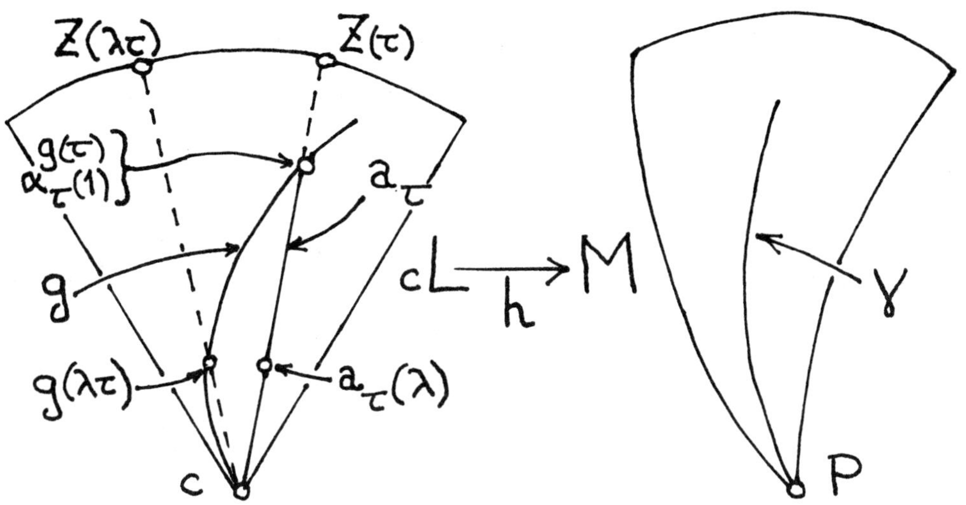

Figure 3

$$= |Dh^{-1}\cdot\underline{\rho}^{\perp}(\gamma'\tau)|/t(\tau), \quad \text{by (1.6.3.,IV)}$$

$$\leq h^{-1}|\underline{\chi}(\tau)|/t(\tau), \quad \text{by Lemma 2.16.1}$$

$$= 0(\tau^{-1+k}),$$

by Lemmas 2.18.1 and 2.20.2. Hence

$$d^{L}(Z(\lambda\tau),Z(\tau)) = \int_{\lambda\tau}^{\tau} \frac{d}{d\tau^{*}} Z(\tau^{*})\,d\tau^{*}$$
$$= 0(\tau^{k}).$$

Proof of (2).

$$(2.21.3) \quad d^{cL}(c,g(\lambda\tau)) = t(\lambda\tau)$$
$$= \lambda\tau(1 + 0((\lambda\tau)^{2k})), \quad \text{by Lemma 2.18.1}$$
$$= \lambda\tau(1 + 0(\tau^{2k})).$$

Similarly,

$$(2.21.4) \quad d^{cL}(c,a_{\tau}(\lambda)) = \lambda t(\tau)$$
$$= \lambda\tau(1 + 0(\tau^{2k})).$$

By (1.16.6),

$$\begin{aligned} d^{cL}(g(\lambda\tau),a_\tau(\lambda)) &\le \{t(\lambda\tau)^2 + \lambda^2 t(\tau)^2 \\ &\quad - 2t(\lambda\tau)\lambda t(\tau)\cos d^L(Z(\lambda\tau),Z(\tau))\}^{\frac{1}{2}} \\ &= \{(\lambda\tau)^2 O(\tau^{2k})\}^{\frac{1}{2}}, \end{aligned}$$

using (1), (2.21.3), (2.21.4) and the fact that $\cos(\theta) = 1 + O(\theta^2)$. Now (2) follows. q.e.d.

COROLLARY 2.21.5. Let $T > 0$ be given and let γ be a geodesic from P with $|\gamma'| \le T$. Let h be a chart at P and assume that $im(\gamma) \subseteq im(h)$. Set $g = h^{-1}o\gamma$, and for $\sigma > 0$ write $g(\sigma) = (Z(\sigma),t(\sigma))$. Let a_σ stand for $a_{g(\sigma)}$. Then for all $\sigma > 0$ and $\lambda \in (0,1]$:

(1) $d^L(Z(\lambda\sigma),Z(\sigma)) = O((\sigma T)^k)$;

(2) $d^{cL}(g(\lambda\sigma),a_\sigma(\lambda)) = \lambda O((\sigma T)^{1+k})$.

Proof. Set

$$\begin{aligned} \tau(\sigma) &= d^M(P,\gamma\sigma) \\ &= \sigma|\gamma'|, \end{aligned}$$

by (2.7.1). Let γ^* be the reparametrization of γ by arc length $\tau(\sigma)$ and apply Lemma 2.21.1 to γ^*. This shows that the left-hand side of (1) is

$$O(\tau(\sigma)^k) = O((\sigma|\gamma'|)^k) = O((\sigma T)^k).$$

This proves (1), and the proof of (2) is similar. q.e.d.

COROLLARY 2.21.6. Let the notation be that of the previous corollary and set $\alpha_\sigma = h o a_\sigma$. Then

$$d^M(\gamma(\lambda\sigma),\alpha_\sigma(\lambda)) = \lambda O((\sigma T)^{1+k}).$$

Proof. This follows from Corollaries 2.21.5 and 2.16.8. q.e.d.

(2.22) Proof of Proposition 2.13. The portion of γ in M^o is C^∞ by Proposition 2.17.3; so by Corollary 2.16.5, $h^{-1}o\gamma$ is C^∞ in cL (in the sense of (2.2)). Write $h^{-1}o\gamma(\sigma) = (Z(\sigma),t(\sigma))$ for $\sigma > 0$. Then $\lim_{\sigma\to 0^+} Z(\sigma)$ exists by Corollary 2.21.5 with $T = |\gamma'|$; and $\lim_{\sigma\to 0^+} t(\sigma)/\sigma = |\gamma'|$ by Corollary

2.18.7. Thus $h^{-1}o\gamma$ is C^1 at c, which is what it means to say that γ is C^1 at P with respect to h. q.e.d.

LEMMA 2.23. Let $T > 0$ be given, and for $i = 1,2$ let γ_i be a geodesic from P with domain [0,1] such that $|\gamma_i'| \leq T$. Let h be a chart at P and assume that $\mathrm{im}(\gamma_i) \subseteq \mathrm{im}(h)$. Set $g_i = h^{-1}o\gamma_i$. Then for $\sigma \in [0,1]$,

$$d^{cL}(g_1(\sigma),g_2(\sigma)) = \sigma\{d^{cL}(g_1(1),g_2(1)) + O(T^{1+k})\}.$$

Proof. Set $a_i = a_{g_i(1)}$; then $a_i(1) = g_i(1)$. By Corollary 2.21.5 with σ replaced by 1 and λ by σ,

$$d^{cL}(g_i(\sigma),a_i(\sigma)) = \sigma O(T^{1+k}).$$

By (1.16.6),

$$\begin{aligned} d^{cL}(a_1(\sigma),a_2(\sigma)) &= \sigma\ d^{cL}(a_1(1),a_2(1)) \\ &= \sigma\ d^{cL}(g_1(1),g_2(1)). \end{aligned}$$

The lemma now follows from the triangle inequality. q.e.d.

COROLLARY 2.23.1. Let γ_1 and γ_2 be as in Lemma 2.23. Then for all sufficiently small σ,

$$d^M(\gamma_1(\sigma),\gamma_2(\sigma)) \leq Hh^{-1}\sigma\ d^M(\gamma_1(1),\gamma_2(1)) + O(T^{1+k}).$$

Proof. This follows from Lemma 2.23 and Corollary 2.16.8. q.e.d.

(2.24) Proof of Proposition 2.14. Using (1.16.10) I may assume that $\mathrm{im}(h) \subseteq \mathrm{im}(h^*)$. Set $\underline{h} = h^{*-1}oh$: $cL \to cL^*$. Write

$$\underline{h}(Z,t) = (Z^*(Z,t),t^*(Z,t)) = (Z^*,t^*),$$

and for $t > 0$ define $\underline{h}_t$: $L \to L^*$ by $\underline{h}_t(Z) = Z^*(Z,t)$. Let b be a path from c in cL, and set $b^* = \underline{h}ob$. Write

$$b(\sigma) = (Z\sigma,t\sigma)\ ,\ b^*(\sigma) = (Z^*\sigma,t^*\sigma).$$

It follows from Corollary 2.16.5 that b^* is C^1 if and only if b is C^1.

Therefore it is enough to show that if b is C^1 at c, then so is b*. So assume that

(2.24.1) $\lim_{\sigma\to 0^+} (Z\sigma, t(\sigma)/\sigma) = (Z0, b)$, say, exists in $c^\infty L$.

I shall prove (2.24.2) and (2.24.3):

(2.24.2) $t^*(\sigma) = \sigma b + o(\sigma;\ \text{fixed } b)$;

(2.24.3) $\underline{h}_t$ converges uniformly as $t \to 0^+$.

Assuming these two assertions for the moment, set $\underline{h}_0 = \lim_{t\to 0^+} \underline{h}_t$. Then it follows that

$$\lim_{\sigma\to 0^+} (Z^*\sigma, t^*(\sigma)/\sigma) = (\underline{h}_0 \circ Z(0), b)$$

exists in $c^\infty L^*$. Proposition 2.14 now follows.

Proof of (2.24.2). Set $\tau(\sigma) = d^M(P, h\circ b(\sigma))$. By (2.24.1),

$$t(\sigma) = \sigma b + o(\sigma;\ \text{fixed } b).$$

Hence

$$\tau(\sigma) = t(\sigma)(1 + O((t\sigma)^{2k})), \quad \text{by Lemma 2.18.2}$$
$$= \sigma b + o(\sigma;\ \text{fixed } b).$$

Since $h\circ b(\sigma) = h^*\circ b^*(\sigma)$, Lemma 2.18.1 gives

$$t^*(\sigma) = \tau(\sigma)(1 + O((\tau\sigma)^{2k}))$$
$$= \sigma b + o(\sigma;\ \text{fixed } b).$$

Proof of (2.24.3). Let $\alpha_{Z,t}$ be as in (2.2.5) with $t \le 1$, and let $\alpha^*_{Z^*,t^*}$ be defined similarly, where $(Z^*, t^*) = \underline{h}(Z,t)$ as above, so that $Z^* = \underline{h}_t(Z)$. (See Figure 4.)

Observe that for $\lambda \in [0,1]$,

(2.24.4) $h^{*-1} \circ \alpha_{Z,t}(\lambda) = \underline{h} \circ a_{Z,t}(\lambda)$

$$= \underline{h}(Z, \lambda t);$$

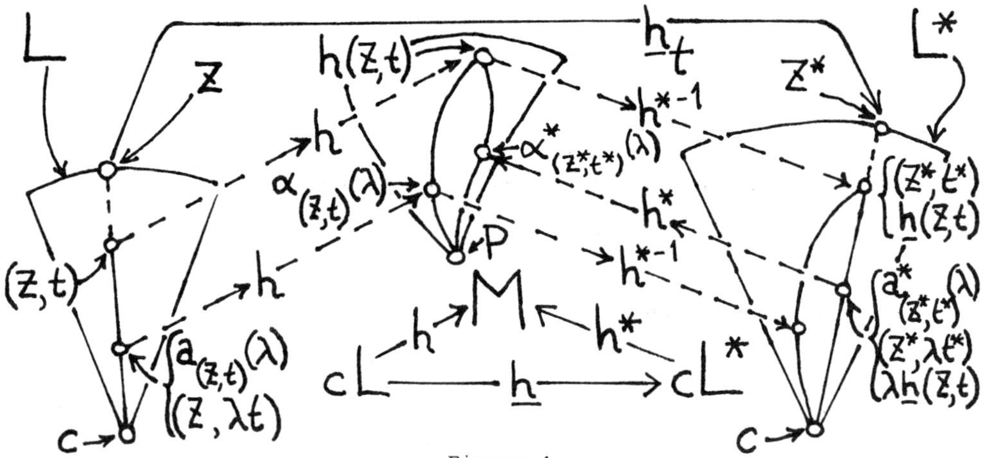

Figure 4

and

$$(2.24.5)\quad h^{*-1}\circ\alpha^*_{Z^*,t^*}(\lambda) = a^*_{Z^*,t^*}(\lambda)$$
$$= (Z^*,\lambda t^*)$$
$$= \lambda\underline{h}(Z,t).$$

Let γ be a geodesic in M from P to $h(Z,t) = h^*(Z^*,t^*)$ parametrized by arc length on $[0,\tau(t)]$, where $\tau(t) = d^M(P,h(Z,t))$. By Corollary 2.21.6 with $T = |\gamma'| = 1$ and $\sigma = \tau(t)$,

$$d^M(\alpha_{Z,t}\lambda,\gamma\lambda) = \lambda O((\tau t)^{1+k}),$$

for all $\lambda \in [0,1]$. Similarly

$$d^M(\alpha^*_{Z^*,t^*}\lambda,\gamma\lambda) = \lambda O((\tau t)^{1+k}).$$

Combining these estimates with Lemma 2.18.2 gives

$$(2.24.6)\quad d^M(\alpha_{Z,t}\lambda,\alpha^*_{Z^*,t^*}\lambda) = \lambda O(t^{1+k}).$$

Therefore

$$(2.24.7)\quad d^{cL^*}(\underline{h}(Z,\lambda t),\lambda\underline{h}(Z,t)) = d^{cL^*}(h^{*-1}\circ\alpha_{Z,t}\lambda, h^{*-1}\circ\alpha^*_{Z^*,t^*}\lambda),$$

by (2.24.4) and (2.24.5)

$$= \lambda O(t^{1+k}),$$

by (2.24.6) and Corollary 2.16.8

$$= \lambda t O(t^k).$$

Now

$$t^*(Z,t) = \tau(t)(1 + O((\tau t)^{2k})), \quad \text{by Lemma 2.18.1 for } h^*$$
$$= t(1 + O(t^{2k})),$$

by Lemma 2.18.2 for h. Hence

$$d^{cL^*}(c,\lambda \underline{h}(Z,t)) = \lambda t^*(Z,t)$$
$$= \lambda t(1 + O(t^{2k})).$$

Similarly

$$d^{cL^*}(c,\underline{h}(Z,\lambda t)) = t^*(Z,\lambda t)$$
$$= \lambda t(1 + O((\lambda t)^{2k}))$$
$$= \lambda t(1 + O(t^{2k})).$$

Comparing these estimates to (2.24.7) shows that once t is small enough, independently of Z,

$$d^{cL^*}(\underline{h}(Z,\lambda t),\lambda \underline{h}(Z,t)) < \min\{d^{cL^*}(c,\underline{h}(Z,\lambda t)),d^{cL^*}(c,\lambda \underline{h}(Z,t))\}.$$

Let d* stand for the left-hand side of this inequality. By (1.16.6),

$$d^{L^*}(Z^*(Z,\lambda t),Z^*(Z,t)) = \cos^{-1}\left\{\frac{t^*(Z,\lambda t)^2 + \lambda^2 t^*(Z,t)^2 - d^{*2}}{2t^*(Z,\lambda t)\lambda t^*(Z,t)}\right\}$$
$$= \cos^{-1}[1 + O(t^{2k})]$$
$$= O(t^k).$$

(De l'Hôpital's rule shows that

$$\cos^{-1}(1 - u) - (2u)^{\frac{1}{2}} = o(u).)$$

That is,

$$d^{L^*}(\underline{h}_{\lambda t}(Z),\underline{h}_t(Z)) = O(t^k).$$

This proves (2.24.3), and with it, Proposition 2.14. q.e.d.

The Metric Tangent Space

(2.25) Definitions. Let β_1 and β_2 be paths from P in M which are C^1 at P (see (2.11)). They are called equivalent if

$$d^M(\beta_1\sigma,\beta_2\sigma) = o(\sigma;\ \text{fixed } \beta_1 \text{ and } \beta_2).$$

The equivalence class of β will be denoted $\hat{\beta}$, and the class of the constant path $\beta(\sigma) \equiv P$ will be denoted $\hat{P}$.

Now let β_1 and β_2 be C^1 at P, not necessarily equivalent. Then

$$\begin{aligned} d^M(\beta_1\sigma,\beta_2\sigma) &\le L(\beta_1 \restriction [0,\sigma]) + L(\beta_2 \restriction [0,\sigma]) \\ &= O(\sigma;\ \text{fixed } \beta_1 \text{ and } \beta_2), \end{aligned}$$

by (2.4.2). Hence

$$\limsup_{\sigma\to 0^+} d^M(\beta_1\sigma,\beta_2\sigma)/\sigma \text{ exists and is finite.}$$

Since this quantity depends only on the equivalence classes of β_1 and β_2, I may define

(2.25.1) $$d^T(\hat{\beta}_1,\hat{\beta}_2) = \limsup_{\sigma\to 0^+} d^M(\beta_1\sigma,\beta_2\sigma)/\sigma.$$

Then d^T is a metric on

$$T_PM = \{\hat{\beta} \text{ for } \beta \text{ a path } C^1 \text{ at } P\}.$$

This metric space is called the tangent metric space of M at P. Set

(2.25.2) $|\hat{\beta}| = d^T(\hat{P},\hat{\beta})$.

Note that

$$d^T(\hat{\beta}_1,\hat{\beta}_2) \le |\hat{\beta}_1| + |\hat{\beta}_2|.$$

When γ is a geodesic from P in M, then by (2.7.1) and Proposition 2.13,

(2.25.3) $|\hat{\gamma}| = |\gamma'|$.

(2.25.4) Remark. It will be shown (Proposition 6.10) that under the further hypotheses VI and VII of (1.6.3),

$$d^T(\hat{\beta}_1, \hat{\beta}_2) = \lim_{\sigma \to 0^+} d^M(\beta_1\sigma, \beta_2\sigma)/\sigma;$$

and also (Theorem 6.15) that d^T is an intrinsic metric in the sense of (2.5.5).

(2.25.5) Example. Let M be a metric cone, cL. Then $T_c(cL)$ can be identified isometrically with cL under the correspondence

$$\hat{b} \leftrightarrow \lim_{\sigma \to 0^+} (Z\sigma, t(\sigma)/\sigma),$$

where $b(\sigma) = (Z\sigma, t(\sigma))$.

THEOREM 2.26. (1) $T_P M$ is an infinite cone with vertex P. (2) Let $h\colon\ cL \to M$ be a chart at P; then h induces a conical homeomorphism

$$T_P h\colon\ c^\infty L \to T_P M.$$

Proof of (1). Given $r > 0$, define $r\colon\ T_P M \to T_P M$ as follows. Let β be a path which is C^1 at P with domain $[0, \sigma_1]$, and define $r(\beta)$ on $[0, \sigma_1/r]$ by

$$r(\beta)(\sigma) = \beta(r\sigma).$$

If β_1 is equivalent to β, then $r(\beta)$ and $r(\beta_1)$ are equivalent; hence I may define

$$r(\hat{\beta}) = \widehat{r(\beta)}.$$

I omit the rest of the proof of (1), which consists of verifying (1.16.8).

Proof of (2). Let b be a path in cL which is C^1 at c; then $\beta = h \circ b$ is C^1 at P. It follows from Corollary 2.16.8 that if b_1 is equivalent to b, then β and $h \circ b_1$ are equivalent. Hence the rule $\hat{b} \to \hat{\beta}$ gives a well-defined map

$$T_P h\colon\ T_c(cL) \to T_P M;$$

in view of (2.25.5), this is the required map.

$T_P h$ is surjective by definition (2.7). Now let $\beta_i = h\circ b_i$ for $i = 1,2$; then it follows from Corollary 2.16.8 that

$$(2.26.3)\quad k d^{c^\infty L}(\hat{b}_1,\hat{b}_2) \leq d^T(\hat{\beta}_1,\hat{\beta}_2) \leq d^{c^\infty L}(\hat{b}_1,\hat{b}_2).$$

This shows that $T_P h$ is one-to-one and bicontinuous. $T_P h$ respects the action (1.16.4) of $\mathbb{R}_+$ on $c^\infty L$ and on $T_P M$. Hence to show $T_P h$ is conical it is enough to show that when $|\hat{b}| = 1$, then $|\hat{\beta}| = 1$ also. Write $\hat{b} = (Z,1) \in c^\infty L$; then b is equivalent to $a_{Z,1}$, which is an integral curve of $\underline{\partial}_t$. Hence $\alpha_{Z,1}$ is a representative of $\hat{\beta}$ and an integral curve of $\underline{\rho}$. Now

$$\begin{aligned} |\hat{\beta}| &= d^T(\hat{P},\hat{\beta}) \\ &= \limsup_{t\to 0^+} d^M(P,\alpha_{Z,1}t)/t \\ &= \limsup_{t\to 0^+} t(1 + 0(t^{2k}))/t, \quad \text{by Lemma 2.18.2} \\ &= 1. \end{aligned}$$

q.e.d.

COROLLARY 2.26.4. As a topological space, L is determined by M. That is, if $h: cL \to M$ and $h^*: cL^* \to M$ are charts at P, then $(T_P h^*)^{-1} \circ T_P h: L \to L^*$ is a homeomorphism.

(2.27) Example. Let M be the real cusp $\{x^3 - y^2 = 0\} \subseteq \mathbb{R}^2$ and let $P = (0,0)$. The Whitney tangent cone of M at P is $\{y^2 = 0\}$, "two copies" of the x-axis. As an extrinsic tangent cone of M in $\mathbb{R}^2$ at P one might use the non-negative x-axis. $T_P M$ is intermediate between these two cones, being two copies of the non-negative x-axis identified at the origin. Thus $T_P M$ is isometric to $\mathbb{R}$, reflecting the fact that the cusp at P is not intrinsic to M. The metric tangent space of the complex form of this example is more complicated; see Corollary 7.18.4.

(2.28) Example. Let M be the surface in $\mathbb{R}^3$ obtained by rotating Example 2.27 about the x-axis; in cylindrical coordinates, $M = \{(r,\theta,z)$ such that $r^2 = z^3\}$. Then there is no chart at the origin, P, such that I -- V of

(1.6.3) hold.

Proof. If such a chart existed then T_PM would be defined and homeomorphic to M near P (compare Remark 2.1); but I shall now outline a proof that $T_PM = \mathbb{R}_+ \cup \{0\}$. Let $(z^{3/2},\theta,z) \in M$ be abbreviated to $[\theta,z]$. For fixed θ, the curve $\beta(z) = [\theta,z]$, when reparametrized by arc length, is a weak geodesic from P by symmetry of M, and is in fact a geodesic because M^o has negative curvature. It follows that

$$d^M(P,[\theta,z]) = z + o(z).$$

Also for $z_1 \leq z_2$,

$$d^M([\theta,z_1],[\theta,z_2]) = z_2 - z_1 + o(z_2).$$

Let $0 \leq \theta_1 \leq \theta_2 \leq 2\pi$, and set $\alpha_z(\theta) = [\theta,z]$ on the domain $[\theta_1,\theta_2]$. Then

$$\begin{aligned} d^M([\theta_1,z],[\theta_2,z]) &\leq L(\alpha_z) \\ &= (\theta_2 - \theta_1)z^{3/2} \\ &= o(z). \end{aligned}$$

Hence for arbitrary $[\theta_1,z_1]$ and $[\theta_2,z_2]$,

$$d^M([\theta_1,z_1],[\theta_2,z_2]) \leq z_2 - z_1 + o(\max(z_1,z_2)).$$

Now for $i = 1,2$, let β_i be a path C^1 at P such that $|\hat{\beta}_i| = 1$, and write $\beta_i(\sigma) = [\theta_i(\sigma),z_i(\sigma)]$. Then

$$d^M(P,\beta_i(\sigma)) = \sigma + o(\sigma),$$

so

$$z_i(\sigma) = \sigma + o(\sigma).$$

Hence

$$d^M(\beta_1\sigma,\beta_2\sigma) = o(\sigma);$$

so β_1 and β_2 are equivalent. That is, the base of the cone T_PM is a single point, establishing the contradiction claimed above.

CHAPTER 3: THE ROLE OF AN UPPER BOUND ON SECTIONAL CURVATURE

Introduction

In this chapter property (1.2) of an exponential map at P is proved: whenever X is sufficiently close to P, then there is a unique geodesic γ_X from P to X with domain [0,1] (Theorem 3.3). By Proposition 2.13, γ_X is C^1 at P, so the rule $X \to \hat{\gamma_X} \in T_PM$ defines a map near P, e_P^{-1}: $M \to T_PM$, called the "inverse-exponential" map. To prove (1.2) one surely needs some upper estimate on the sectional curvature function K^M to ensure that within some neighbourhood of P no conjugate point to P can occur along any geodesic from P. Hypothesis (1.6.3,VI) accomplishes the purpose but is stronger than necessary (see Example 3.2).

The proof of Theorem 3.3 is in outline the following. Let γ_X and γ_* be geodesics from P to X, both parametrized by [0,1]; and set $\tau(X) = d^M(P,X)$. Using Corollary 2.18.1 once can show that given $\underline{r} > 0$ there exists $\underline{\tau}^{\#} > 0$ such that, so long as $\tau(X) < \underline{\tau}^{\#}$, $d^M(\gamma_X\lambda,\gamma_*\lambda) < \lambda\underline{r}\tau(X)$ for all $\lambda \in (0,1]$ (see (3.13)). Let β^λ be a geodesic from $\gamma_X(\lambda)$ to $\gamma_*(\lambda)$. Then, provided $\underline{r} < 1$, $\text{im}(\beta^\lambda) \subseteq B(X,M;\tau X) \subseteq M^o$. Now assume $\tau(X) < \pi/(2K^{\frac{1}{2}})$, where K is as in (1.6.3,VI). Rauch's comparison theorem, comparing M^o to an n-sphere of radius $K^{-\frac{1}{2}}$, implies that the exponential map e_X at X restricts to an immersion

$$e_X\colon\ B(X,T_XM;\tau X) \to B(X,M;\tau X),$$

and that $\|De_X\|^{-1}$ is bounded on $B(X,T_XM;\tau X)$. There is a natural lifting $\tilde{\gamma}_X$ of γ_X to T_XM, namely $\tilde{\gamma}_X(\lambda) = (1-\lambda)\gamma_X{}'(1)$; and similarly for γ_*. If it can be proved that each β is <u>unique</u>, then the family β^λ is continuous in λ; and a covering homotopy argument shows that there is a unique lifting $\tilde{\beta}^\lambda$ of β^λ to a path from $\tilde{\gamma}_X(\lambda)$ to $\tilde{\gamma}_*(\lambda)$ (Lemma 3.15). The bound on $\|De_X\|^{-1}$ implies that $L(\tilde{\beta}^\lambda) = O(L\beta^\lambda)$, uniformly in λ. But

$$L(\beta^\lambda) = d^M(\gamma_X\lambda,\gamma_*\lambda) \to 0 \text{ as } \lambda \to 0,$$

while

$$\begin{aligned} L(\tilde{\beta}^{\lambda}) &= |\tilde{\gamma}_X(1-\lambda) - \tilde{\gamma}_*(1-\lambda)| \\ &= (1-\lambda)|\gamma_X'(1) - \gamma_*'(1)|. \end{aligned}$$

This can happen only if $\gamma_X'(1) = \gamma_*'(1)$; but now γ_X and γ_* are both weak geodesics in M^o which have the same final conditions at X, and are therefore equal. To prove Theorem 3.3 it remains to show that β^λ is unique for each λ. This follows by taking $X = \gamma_X(\lambda)$ in the following result (Proposition 3.7): if $\underline{r}^*$ and $\underline{\tau}^*$ are chosen small enough ($\underline{r}$ and $\underline{\tau}^\#$ are chosen still smaller), then whenever $\tau(X) < \underline{\tau}^*$, e_X is a diffeomorphism on $B(X, T_XM; \underline{r}^*\tau X)$. The difficult step consists of showing that e_X is one-to-one near X, which reduces to proving that the space Ω of paths of small energy from X to a fixed nearby Y is connected (see (3.11)). Let h: $cL \to M$ be a chart at P. It is easy to prove the analogue of Proposition 3.7 for cL (Lemma 3.8). By results of Palais [12], the corresponding path space Ω_1 is contractible (Lemma 3.10). If Ω were not connected, one could find geodesics β_0 and β_1 from X to Y in different components of Ω. Now $b_0 = h^{-1} \circ \beta_0$ and $b_1 = h^{-1} \circ \beta_1$ are homotopic in Ω_1; and a homotopy b_t between them can be chosen to consist of paths so sparing in energy that $\beta_t = h \circ b_t$ is a homotopy in Ω (Lemma 3.10); this is a contradiction, which proves Proposition 3.7, and with it, Theorem 3.3.

(3.1) Throughout this chapter I shall assume (1.6.1), (1.6.2) and I -- VI of (1.6.3).

(3.1.1) Caution. I speak of an "inverse-exponential" map at P and denote it e_P^{-1}, but it is not assumed that e_P^{-1} is the inverse of any map. In fact the proofs in Chapter 6 that e_P^{-1} is continuous and has as image a neighbourhood of P in T_PM require only the hypotheses of (3.1); but to prove e_P^{-1} one-to-one seems to require (1.6.3,VII).

(3.2) Example. Let L satisfy (1.6.2) and give L a Riemannian metric. Then cL itself, with the identity map as chart, satisfies (1.6.1), (1.6.2) and I -- V and VII of (1.6.3), but not necessarily VI. In fact (1.6.3,VI) holds if and only if

(3.2.1) $K^L < 1$ everywhere on L.

Proof. Let II be the second fundamental form of $L = L \times 1$ in $c^\infty L$. Let Π be a 2-plane tangent to L at Z, and let v,w be an orthonormal basis of Π. Then (see Hicks [8, p. 78]).

(3.2.2) $K^L(\Pi) = K^{c^\infty L}(\Pi) + II(v,v)II(w,w) - II(v,w)^2.$

Now $\underline{\partial}_t$ is a unit normal vector field to L; and it follows that $II(v,w) = \langle v,w \rangle$. Thus (3.2.2) reduces to

(3.2.3) $K^{c^\infty L}(\Pi) = K^L(\Pi) - 1.$

The map i_t (see (1.16.2)) induces from Π a 2-plane $Di_t \cdot \Pi$ tangent to $L \times t$ at $Z \times t$, for each $t > 0$. Then, using (1.16.6),

(3.2.4) $K^{c^\infty L}(Di_t \cdot \Pi) = t^{-2} K^{c^\infty L}(\Pi).$

Finally

(3.2.5) $K^{c^\infty L}(\Pi') = 0$, whenever $\underline{\partial}_t \in \Pi'$, everywhere on $(c^\infty L)^0$.

It follows from (3.2.3) -- (3.2.5) that (3.2.1) holds. q.e.d.

On the other hand, as noted in Remark 1.7.2, $c^\infty L$ trivially has an exponential map at c for any L as above, whatever the sectional curvature K^L may be. This suggests that (1.6.3,VI) can be weakened. (See also Remark 4.10.1.) Such a result would be particularly welcome if it applied to complex analytic hypersurfaces with unbranched isolated singular points; see Example 7.24 and Conjecture 7.24.1.

It is easy to arrange that (3.2.1) hold. For since L is compact K^L is at all events bounded above; and changing the given metric on L by a suitable scale factor yields (3.2.1). However in the proof of Lemma 3.10 it is necessary that L be a submanifold of the unit sphere $S^{N-1} \subseteq \mathbb{R}^N$ for some N. In this case (3.2.1) cannot be guaranteed, but I may assume instead:

(3.2.6) $K^L < K$ (of (1.6.3,VI)) everywhere on $L \subseteq S^{N-1}$.

The Inverse-Exponential Map at P

The main purpose of this chapter is to prove:

THEOREM 3.3. Assume that (1.6.1), (1.6.2) and I -- VI of (1.6.3) hold. Then there exists $\underline{\tau}^{\#} \in (0,1)$ such that for every $X \in B(P,M;\underline{\tau}^{\#})$ there is a unique geodesic from P to X with domain [0,1]. Thus property (1.2) of an exponential map at P is valid.

The proof is completed in (3.16). A suitable choice of $\underline{\tau}^{\#}$ is given in (3.13). (3.4) and (3.5) are immediate consequences of the theorem.

(3.4). Notation. For any $X \in M$, set $\tau(X) = d^M(P,X)$ as in (2.15.4). When $\tau(X) < \underline{\tau}^{\#}$, let γ_X be the geodesic from P to X with domain [0,1]; and let γ^X be the same geodesic parametrized by arc length, so γ^X has domain $[0,\tau(X)]$. In view of (2.7.1) and (2.25.3).

(3.4.1) $|\hat{\gamma}^X| = 1,\ \hat{\gamma}_X = \tau(X)\hat{\gamma}^X,\ |\hat{\gamma}_X| = \tau(X).$

Let $Y = \gamma_X(\lambda)$, for $\lambda \in [0,1]$. Then $\gamma_Y(\sigma) = \gamma_X(\lambda\sigma)$; hence

(3.4.2) $\hat{\gamma}_Y = \lambda\hat{\gamma}_X.$

On $B^o(P,M;\underline{\tau}^{\#})$ define vector fields $\underline{\Gamma}$ and $\underline{\partial}_\tau$ by

(3.4.3) $\underline{\Gamma}(X) = \gamma_X'(1)$, and

(3.4.4) $\underline{\partial}_\tau(X) = |\underline{\Gamma}(X)|^{-1}\underline{\ }(X) = \gamma^{X\prime}(\tau X).$

The latter of these fields is called the unit radial vector field from P.

(3.5) The inverse-exponential map e_P^{-1}: $B(P,M;\underline{\tau}^{\#}) \to T_PM$ is defined by

$$e_P^{-1}(X) = \hat{\gamma}_X,$$

where $\underline{\tau}^{\#}$ is as in Theorem 3.3 and (3.13) and γ_X as in (3.4).

(3.6) Notation. Let $X \in M^o$ and let e_X denote the exponential map at X. The natural domain $\tilde{U}_X$ of e_X consists of those $\tilde{Y} \in T_XM$ such that there exists a weak geodesic β: $[0,1] \to M^o$ satisfying $\beta(0) = X$ and $\beta'(0) = \tilde{Y}$.

I now embark upon the proof of Theorem 3.3.

Proposition 3.7. There exist $\underline{r}^* \in (0,1)$ and $\underline{\tau}^* > 0$ such that whenever $X \in B(P,M;\underline{\tau}^*)$, then:

(1) $B(X,T_XM;\tau X) \subseteq \tilde{U}_X$;

(2) Let $\tilde{Y} \in \tilde{U}_X \cap B(X,T_XM;\underline{\tau}^*)$; then $\tilde{Y}$ is a regular point of e_X, and

$$|De_X(\tilde{Y})\cdot\tilde{\psi}| \geq (1-|Y|^2/6)|\tilde{\psi}| \geq |\tilde{\psi}|/2, \text{ for } \tilde{\psi} \in T_{\tilde{Y}}(T_XM),$$

$$|D(e_X^{-1})(Y)\cdot\psi| \leq (1+|\tilde{Y}|^2/6)|\psi| \leq (3/2)|\psi|, \text{ for } \psi \in T_YM,$$

where $Y = e_X(\tilde{Y})$;

(3) e_X is a diffeomorphism on $B(X,T_XM;\underline{r}^*\tau(X))$.

Proof. The proof incorporates Lemmas 3.8 and 3.10. Suitable choices of $\underline{r}^*$ and $\underline{\tau}^*$ are given in (3.9).

Proof of (1). Let h be a chart at P. Choose $\underline{\tau}^*$ such that

(3.7.4) $B(P,M;2\underline{\tau}^*) \subseteq \text{im}(h)$.

Then whenever $X \in B(P,M;\underline{\tau}^*)$,

$$B(X,M;\tau X) \subseteq \text{im}(h).$$

Since im(h) is complete by Corollary 2.16.9, assertion (1) follows from the Hopf-Rinow theorem (see GKM [6, p. 166]).

Proof of (2). In view of (1.6.3,VI), Rauch's comparison theorem may be applied to M^o and a Euclidean n-sphere of radius $K^{-\frac{1}{2}}$, K being as in (1.6.3, VI). By an elementary consequence of that theorem (see GKM [6, p. 179]), whenever

(3.7.5) $\underline{\tau}^* < \pi/(2K^{\frac{1}{2}})$

and whenever $\tilde{Y} \neq 0$ is as in (2), then

$$|De_X(\tilde{Y})\cdot\tilde{\psi}| > |\tilde{\psi}|\sin(|\tilde{Y}|K^{\frac{1}{2}})/(|\tilde{Y}|K^{\frac{1}{2}}).$$

The first estimate of (2) now follows from (1.17.1) when $\tilde{Y} \neq 0$ and from the the fact that $De_X(X) = \text{id}_{T_XM}$ when $\tilde{Y} = 0$. Therefore $\tilde{Y}$ is a regular point of e_X; and now the second estimate of (2) follows from the first one and (1.17.2).

Proof of (3). Let $h\colon cL \to M$ be a chart at P, and assume that (3.2.6) holds. The next lemma proves an analogue of Proposition 3.7 for a metric cone, which need not, however, satisfy (1.6.3,VI).

LEMMA 3.8. Under the assumptions above, there exists $r_1 \in (0,1)$ such that for all $(Z,t) \in (c^\infty L)^o$:

(1) $B(r_1) = B((Z,t), T_{(Z,t)}c^\infty L;\ tr_1) \subseteq \tilde{U}_{(Z,t)}$;

(2) Whenever $\tilde{Y} \in B(r_1)$ and $\tilde{v} \in T_{\tilde{Y}}T_{(Z,t)}c^\infty L$, then

$$|De_{(Z,t)}(\tilde{Y})\cdot\tilde{v}| \geq (1 - |\tilde{Y}|^2/6)|\tilde{v}| \geq |\tilde{v}|/2;$$

(3) $e_{(Z,t)}$ is a diffeomorphism on $B(r_1)$.

Proof. Because the metric on $c^\infty L$ is conical in the sense of (1.16.8), it is sufficient to find r_1 such that (1), (2) and (3) hold when $t = 1$. Pick r' such that

$$(3.8.4)\qquad 0 < r' < \begin{cases} 1, & \text{if } K \in (0,1], \\ 1 - (1 - K^{-1})^{1/2}, & \text{if } K > 1. \end{cases}$$

Then for any $r \in (0,r']$, and for any $(Z^*,t^*) \in B(r)$,

$$(3.8.5)\qquad t^* > 1-r > \begin{cases} 0, & \text{if } K \in (0,1], \\ (1-K^{-1})^{1/2}, & \text{if } K > 1. \end{cases}$$

Let $\Pi^* \subseteq T_{(Z^*,t^*)}c^\infty L$ be any tangent 2-plane, and let $\Pi \subseteq T_{(Z^*,1)}c^\infty L$ be the corresponding 2-plane (the image of Π^* under the dilation by $1/t^*$). Then the sectional curvature (see (1.15.6))

$$K^{c^\infty L}(\Pi^*) = (t^*)^{-2}K^{c^\infty L}(\Pi), \qquad \text{by (3.2.3)}$$
$$\begin{cases} \leq 0 \quad , & \text{if } K \in (0,1], \\ < (1-K)^{-1}(K-1) = \quad , & \text{if } K > 1, \end{cases}$$

by (3.2.2), (3.2.4), (3.2.6) and (3.8.5). That is,

(3.8.6) $K^{c^\infty L} < K$ on $B(r)$ whenever r is as above.

Let $r(Z)$ be the radius of convexity of $c^\infty L$ at $(Z,1)$ (see GKM [6, p. 162]). Then $r(Z)$ varies continuously in Z; hence

(3.8.7) $r'' = \inf\{r(Z) \text{ for } Z \in L\}$ is positive.

Pick r_1 such that, with r' as in (3.8.4) and r'' as in (3.8.7),

(3.8.8) $0 < r_1 < \min\{r', \pi/(2K^{1/2}), r''\}$.

Then since $r_1 < r' < 1$, $B(r_1) \subseteq (c^\infty L)^o$; hence (1) holds. In view of (3.8.6), the proof of Proposition 3.7.2 applies, since $r_1 < \pi/(2K^{1/2})$, to show (2). Therefore $e_{(Z,1)}$ is an immersion on $B(r_1)$. By choice of r'', and since $r_1 < r''$, $e_{(Z,1)}$ is one-to-one on $B(r_1)$. This proves (3).

q.e.d.

(3.9) <u>Proof of Proposition 3.7.3, continued</u>. Let H and h be as in Lemma 2.16.1. Choose

(3.9.1) $\underline{r}^* < \min\{1/2, hr_1/H\}$,

where r_1 is as in (3.8.8); and choose

(3.9.2) $\underline{\tau}^* < \min\{\pi/(2K^{1/2}), h/2\}$ such that

$$B(P,M;\underline{\tau}^*(1+\underline{r}^*)) \subseteq \mathrm{im}(h).$$

Assume that $\tau(X) < \underline{\tau}^*$. To prove Proposition 3.7.3 it is enough to show, in view of Proposition 3.7.2, that e_X is one-to-one on

$$\tilde{B} = B(X,T_XM;\underline{r}^*\tau(X)).$$

Suppose the contrary, and let $\tilde{Y}_0,\tilde{Y}_1 \in B$ be such that $e_X(\tilde{Y}_0) = e_X(\tilde{Y}_1) = Y$, say. Let β_i, for $i = 0,1$, be the geodesic $\beta_i(\sigma) = e_X(\sigma\tilde{Y}_i)$. By (3.9.2), $\mathrm{im}(\beta_i) \subseteq \mathrm{im}(h)$; so set $b_i = h^{-1}\circ\beta_i$ and write $h^{-1}(X) = (Z,t(X))$. Then by Corollary 2.16.7,

(3.9.3) $E(b_i) \leq h^{-2}E(\beta_i)$

$$= h^{-2}|\ddot{Y}_i|^2$$

$$< (\underline{r}^*\tau(X)/h)^2$$

$$< (r_1\tau(X)/H)^2,$$

by (3.9.1). By Corollary 2.16.8, $t(X) \leq H^{-1}\tau(X)$; hence

$$E(b_i) < (r_1 t(X))^2.$$

Therefore $L(b_i) < r_1 t(X)$, so that

$$\mathrm{im}(b_i) \subseteq B((Z,tX), c^\infty L;\ r_1 t(X))$$

$$= B(r_1),$$

for short.

LEMMA 3.10. There is a homotopy b_t from b_0 to b_1 in $B(r_1)$, with fixed endpoints, such that for every t,

$$E(b_t) \leq \max\{E(b_0),\ E(b_1)\}. \quad \text{(See Figure 5.)}$$

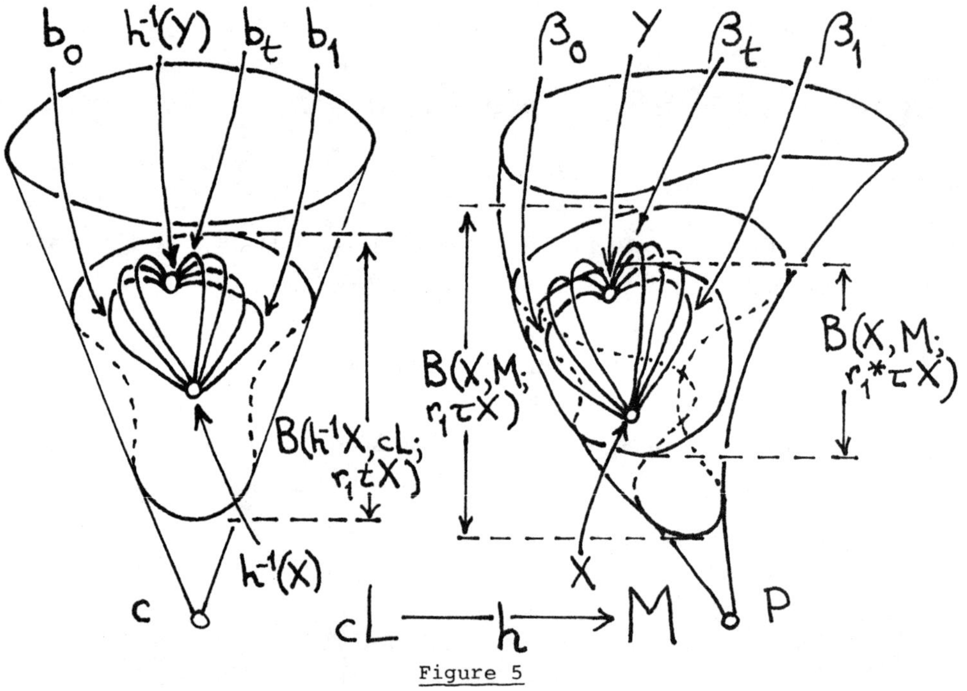

Figure 5

Proof. Because of the assumption (3.2.6) I may assume that $c^{\infty}L$ is isometrically embedded as a cone in some Euclidean space, so that the work of Palais [12] may be applied. Let Ω be the Hilbert manifold of absolutely continuous paths b from $h^{-1}(X)$ to $h^{-1}(Y)$ in $c^{\infty}L$ such that b' is square-summable. Set $\Omega_1 = \{b \in \Omega$ such that $E(b) < (r_1 t(X))^2\}$. For every $b \in \Omega_1$, $L(b) \leq E(b)^{\frac{1}{2}} < r_1 t(X)$; so $\mathrm{im}(b) \subseteq B(r_1)$. By Lemma 3.8.3, $h^{-1}(X)$ and $h^{-1}(Y)$ are not conjugate along any geodesic in Ω_1. Therefore, by Palais [12, Main Theorem and §14], $E\colon \Omega_1 \to \mathbb{R}$ has only non-degenerate critical points. In fact, critical points of E correspond to geodesics from $h^{-1}(X)$ to $h^{-1}(Y)$ and there is only one such geodesic in $B(r_1)$, again by Lemma 3.8.3. It follows that E has only one critical point on Ω_1, which must be a minimum. Hence Ω_1 is contractible. Therefore there exists a path b^*_t from b_0 to b_1 in Ω_1. Now b^*_t can be deformed, keeping its endpoints fixed, downward along the gradient field of E to obtain a path b_t as required. q.e.d.

(3.11) Proof of Proposition 3.7.3, concluded. Set $\beta_t = h \circ b_t$. Then

$$
\begin{aligned}
E(\beta_t) &\leq H^2 E(b_t), && \text{by Corollary 2.16.7}\\
&< H^2(\underline{r}^*\tau(X)/h)^2, && \text{by (3.9.3) and Lemma 3.10}\\
&\leq (r_1\tau(X))^2, && \text{by (3.9.1)}\\
&< \tau(X)^2 && \text{by (3.8.8) and (3.8.4)}\\
&\leq \underline{\tau}^{*2}, && \text{by assumption on } \tau(X).
\end{aligned}
$$

Let $\bar{\Omega}$ be the Hilbert manifold of absolutely continuous paths β from X to Y in M with domain [0,1] such that β' is square-summable and $E(\beta) < (r_1\tau(X))^2$. All such paths lie in $B = B(X,M;r_1\tau(X))$.

Now the closure of B lies in M^o (again by (3.8.8) and (3.8.4)), and so B is contained in a compact, C^{∞} Riemannian manifold with smooth boundary. By a theorem of Nash [11], B can be isometrically embedded in some Euclidean space. Moreover by Proposition 3.7.2 it is impossible to find a geodesic from X to Y in B, of length $< \underline{\tau}^*$, which admits a Jacobi field vanishing at X and at any other point. It now follows from Palais [12, Main Theorem and §14] that the critical points of $E\colon \bar{\Omega} \to \mathbb{R}$ are all non-degenerate, and further, that E has no critical points of index ≥ 1.

Therefore E can have only one critical point in each component of $\bar{\Omega}$. But β_0 and β_1, being geodesics, are critical points of E, and yet β_t is a path between them in $\bar{\Omega}$. This is a contradiction. It follows that e_X is one-to-one on $B(X,T_XM;\underline{r}^*\tau(X))$, which completes the proof of Proposition 3.7.3.

q.e.d.

Theorem 3.3 is a special case of the following result, which will be needed in Chapter 5.

Proposition 3.12. There exists $\underline{\tau}^\# \in (0,1)$ such that whenever $X \in B(P,M;\underline{\tau}^\#)$ the following holds. Let γ_X be a geodesic from P to X and γ_* a weak geodesic from P to X, both having domain $[0,1]$. Assume that for every $\lambda \in (0,1]$,

$$d^M(\gamma_X\lambda,\gamma_*\lambda) < \lambda\underline{r}^*\tau(X)$$

where $\underline{r}^*$ is as in Proposition 3.7 and (3.9.1). Then $\gamma_X = \gamma_*$.

(3.13) Proof of Theorem 3.3 from Proposition 3.12. Let $\underline{r}^*$ and $\underline{\tau}^*$ be as in (3.9). Choose

(3.13.1) $\underline{r} < h\underline{r}^*/8H (\leq \underline{r}^*/8)$.

(3.13.2) Choose $\underline{\tau}^\# < \min\{\underline{\tau}^*,1\}$ such that whenever γ_X and γ_Y are geodesics from P to X and to Y respectively, satisfying $\tau(X) < \underline{\tau}^\#$ and $d^M(X,Y) < \underline{r}\tau(X)$, then

$$d^M(\gamma_X\lambda,\gamma_Y\lambda) < 2\lambda H(d^M(X,Y) + \underline{r}\tau(X))/h,$$

for all $\lambda \in (0,1]$.

It follows from Corollary 2.23.1 that $\underline{\tau}^\#$ can be so chosen. For set $T = \tau(X)(1+\underline{r})$. Then

$$|\gamma_X'| = d^M(P,X) = \tau(X) < T,$$

and

$$\begin{aligned}|\gamma_Y'| &= d^M(P,Y)\\ &\leq d^M(P,X) + d^M(X,Y)\\ &< \tau(X) + \underline{r}\tau(X)\end{aligned}$$

$$= T.$$

So Corollary 2.23.1 applies, to give

$$d^M(\gamma_X\lambda,\gamma_Y\lambda) < \lambda H(d^M(X,Y) + O(T^{1+k}))/h.$$

With the above choice of T,

$$O(T^{1+k}) = \tau(X)(1+\underline{r})^{1+k}O(\underline{\tau}^{\#k}) < \underline{r}\tau(X),$$

once $\underline{\tau}^{\#}$ is small enough. Thus (3.13.2) is feasible.

Now assume that X = Y, so that γ_X and γ_Y are both geodesics from P to X. Then it follows from (3.13.1) and (3.13.2) that the hypotheses of Proposition 3.12 hold; so by that result, $\gamma_X = \gamma_Y$. This proves Theorem 3.3, assuming Proposition 3.12. q.e.d.

(3.14) Proof of Proposition 3.12. The proof incorporates Lemma 3.15 and is concluded in (3.16). By Proposition 3.7.3 there is a unique geodesic β^λ from $\gamma_X(\lambda)$ to $\gamma_*(\lambda)$ in M^o, for each $\lambda \in (0,1]$. The family $\{\beta^\lambda\}$ is uniformly bounded, and it is equicontinuous since $|\beta^{\lambda\prime}| = L(\beta^\lambda)$ is uniformly bounded above. Define

$$\underline{\beta}:\ [0,1]\times(0,1]\to M^o$$

by $\underline{\beta}(\sigma,\lambda) = \beta^\lambda(\sigma)$; then by the Ascoli-Arzela theorem and the uniqueness of each β^λ, $\underline{\beta}$ is continuous. Moreover

$$\begin{aligned} d^M(X,\underline{\beta}(\sigma,\lambda)) &\le d^M(X,\gamma_X\lambda) + L(\beta_\lambda) \\ &< (1-\lambda)\tau(X) + \lambda\underline{r}^*\tau(X) \\ &< \tau(X). \end{aligned}$$

Thus

(3.14.1) $\quad \mathrm{im}(\underline{\beta}) \subseteq B(X,M;\tau(X))$.

LEMMA 3.15. Under the hypotheses of Proposition 3.12 and with the notation above, there exists a unique continuous

$$\underline{\tilde{\beta}}: \quad [0,1] \times (0,1] \to B(X, T_XM; \tau(X))$$

such that:

(1) $e_X \circ \underline{\tilde{\beta}} = \underline{\beta}$;

(2) $\underline{\beta}(0,\lambda) = (\lambda-1)\gamma_X'(1)$.

Moreover $\underline{\tilde{\beta}}$ satisfies

(3) $\underline{\tilde{\beta}}(1,\lambda) = (\lambda-1)\gamma_*'(1)$.

(See Figure 6.)

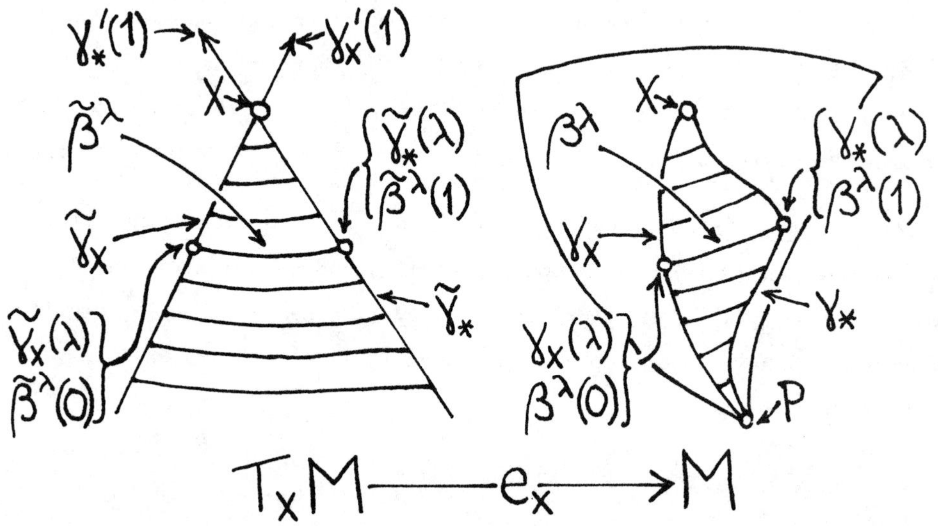

Figure 6

Proof of (1) and (2). I first show that for each λ there exists a unique

$$\tilde{\beta}^\lambda: \quad [0,1] \to B(X, T_XM; \tau(X)) = \tilde{B},$$

say, such that

$$e_X \circ \tilde{\beta}^\lambda = \beta^\lambda, \quad \text{and}$$

$$\tilde{\beta}^\lambda(0) = (\lambda-1)\gamma_X'(1).$$

Let $\sum^\lambda$ be the set of $\underline{\sigma} \in [0,1]$ such that $\tilde{\beta}^\lambda$ can be uniquely defined on

$[0,\underline{\sigma}]$ with the properties above.

Σ^λ is non-empty, for $0 \in \Sigma^\lambda$.

Σ^λ is open. Let $\underline{\sigma} \in \Sigma^\lambda$ with $\underline{\sigma} < 1$; then $[0,\underline{\sigma}] \subseteq \Sigma^\lambda$. Observe that

$$\begin{aligned} |\tilde{\beta}^\lambda(\underline{\sigma})| &= d^M(X,\beta^\lambda(\underline{\sigma})) \\ &< \tau(X), \qquad \text{by (3.14.1)} \\ &< \underline{\tau}^*, \end{aligned}$$

by hypothesis of Proposition 3.12 and (3.13.2). By Proposition 2.7.2, e_X is a diffeomorphism on some neighbourhood of $\tilde{\beta}^\lambda(\underline{\sigma})$ in B. Hence there is a unique extension of $\tilde{\beta}^\lambda$ over $[0,\sigma")$ for some $\sigma" > \underline{\sigma}$. Thus $[0,\sigma") \subseteq \Sigma^\lambda$, which is therefore open.

Σ^λ is closed. Let $[0,\underline{\sigma}) \subseteq \Sigma^\lambda$ and let $\sigma \in [0,\underline{\sigma})$. By Proposition 3.7.2

$$\begin{aligned} |\beta^{\lambda\prime}(\sigma)| &= |De_X(\tilde{\beta}^\lambda\sigma)\cdot\tilde{\beta}^{\lambda\prime}(\sigma)| \\ &\geq |\tilde{\beta}^{\lambda\prime}(\sigma)|/2. \end{aligned}$$

So

$$\begin{aligned} (3.15.4) \qquad |\tilde{\beta}^{\lambda\prime}(\sigma)| &\leq 2|\beta^{\lambda\prime}(\sigma)| \\ &= 2L(\beta^\lambda), \qquad \text{since } \beta^\lambda \text{ is a geodesic} \\ &= 2d^M(\gamma_X\lambda,\gamma_*\lambda) \\ &< 2\lambda\underline{r}^*\tau(X), \end{aligned}$$

by hypothesis of Proposition 3.12. Thus $|\tilde{\beta}^{\lambda\prime}(\sigma)|$ is bounded above uniformly in σ. It follows that $\lim_{\sigma\to\underline{\sigma}^-} \tilde{\beta}^\lambda(\sigma)$ exists in T_XM; and $\tilde{\beta}^\lambda(\underline{\sigma})$ can only be defined as this limit.

From (3.15.4),

$$\begin{aligned} (3.15.5) \qquad (\beta^\lambda \restriction [0,\sigma]) &\leq 2\sigma L(\beta^\lambda) \\ &\leq 2L(\beta^\lambda). \end{aligned}$$

Now

$$
\begin{aligned}
|\tilde\beta^\lambda(\sigma)| &\le |\tilde\beta^\lambda(0)| + L(\tilde\beta^\lambda\restriction[0,\sigma]) \\
&\le (1-\lambda)\tau(X) + 2L(\beta^\lambda), \qquad \text{by (1) and (3.15.5)} \\
&< (1-\lambda)\tau(X) + 2\lambda\underline{r}^*\tau(X), \qquad \text{since } \lambda > 0 \\
&< \tau(X),
\end{aligned}
$$

since $\underline{r}^* < 1/2$ by (3.9.1). Thus $\tilde\beta^\lambda(\sigma) \in B$. Hence $\underline{\sigma} \in \Sigma^\lambda$, proving that Σ^λ is closed.

It follows that $\Sigma^\lambda = [0,1]$, so that $\tilde\beta^\lambda$ is uniquely defined on $[0,1]$ for all $\lambda \in (0,1]$. Set

$$\underline{\tilde\beta}(\sigma,\lambda) = \tilde\beta^\lambda(\sigma).$$

Then (1) and (2) hold, and it remains to show that $\underline{\tilde\beta}$ is continuous. The family $\{\tilde\beta^\lambda\}$ is uniformly bounded, and (3.15.4) shows that it is equicontinuous. It follows from the Ascoli-Arzela theorem and the uniqueness of each $\tilde\beta^\lambda$ that $\underline{\tilde\beta}$ is continuous in λ; in fact, for each $\underline{\lambda}$, $\tilde\beta^\lambda \to \tilde\beta^{\underline{\lambda}}$ uniformly as $\lambda \to \underline{\lambda}$. From this fact and the equicontinuity of $\underline{\tilde\beta}$ in σ it follows that $\underline{\tilde\beta}$ is continuous in (σ,λ).

Proof of (3). Set $\tilde\beta_1(\lambda) = \underline{\tilde\beta}(1,\lambda)$. Then $\tilde\beta_1(1) = X$ and $e_X \circ \tilde\beta_1(\lambda) = \gamma_*(\lambda)$. Since γ_* is a weak geodesic, the path

$$\tilde\beta_1^*(\lambda) = (\lambda-1)\gamma_*'(1)$$

satisfies the same two properties. The argument used to show that each $\tilde\beta^\lambda$ is unique can be applied (in simpler form) to show that $\tilde\beta_1 = \tilde\beta_1^*$, which proves (3). q.e.d.

(3.16) Proof of Proposition 3.12, concluded. For every $\lambda \in (0,1]$,

$$
\begin{aligned}
L(\beta^\lambda) &\ge (1/2)L(\tilde\beta^\lambda), \qquad \text{by (3.15.5)} \\
&\ge (1/2)|(\lambda-1)\gamma_X'(1) - (\lambda-1)\gamma_*'(1)|, \qquad \text{by Lemma 3.15} \\
&= (1/2)(1-\lambda)|\gamma_X'(1) - \gamma_*'(1)|.
\end{aligned}
$$

By hypothesis, however,

$$L(\beta^\lambda) < \lambda\underline{r}^*\tau(X).$$

These inequalities are consistent as $\lambda \to 0$ only if $\gamma_X'(1) - \gamma_*'(1) = 0$. But now $\gamma_X \restriction (0,1]$ and $\gamma_* \restriction (0,1]$ are both weak geodesics in the Riemannian manifold M^o, and they have the same final conditions. Therefore $\gamma_X = \gamma_*$ on $(0,1]$. Thus $\gamma_X = \gamma_*$, which proves Proposition 3.12. q.e.d.

The final lemma of this chapter will be needed in Chapter 5.

LEMMA 3.17. Let $\underline{\tau}^\#$ be as in Theorem 3.3, and let γ be a geodesic from P parametrized by arc length on $[0,\tau]$ with $\tau < \underline{\tau}^\#$. **Then** γ **can be extended to** a geodesic on $[0,\underline{\tau}^\#]$.

Proof. Let

$$\tau^* = \text{l.u.b.}\{\tau' \in [\tau,\underline{\tau}^\#] \text{ such that } \gamma \text{ can be extended to a geodesic on } [0,\tau']\}.$$

Then γ is a geodesic on $[0,\tau^*)$, and $\lim_{\tau'\to\tau^*-}\gamma(\tau')$ exists. Let $\gamma(\tau^*)$ be this limit; then γ is a geodesic on $[0,\tau^*]$. Suppose that $\tau^* < \underline{\tau}^\#$. Since $d^M(P,\gamma\tau^*) = \tau^*$, $\gamma(\tau^*) \in M^o$. So γ can be extended as a weak geodesic (see (1.15.7)) to some $[0,\tau'']$ with $\tau'' \in (\tau^*,\underline{\tau}^\#]$, such that $\gamma\restriction[\tau^*,\tau'']$ is a geodesic. Since

$$d^M(P,\gamma\tau'') \le \tau'' < \underline{\tau}^\#,$$

there is by Theorem 3.3 a unique geodesic γ'' from P to $\gamma(\tau'')$ parametrized by arc length. Set

$$\underline{\tau}'' = d^M(P,\gamma\tau''),$$

so γ'' has domain $[0,\underline{\tau}'']$. (See Figure 7.)

Set $\delta = \tau'' - \tau^*$. Since $\gamma\restriction[\tau^*,\tau'']$ is a geodesic,

$$d^M(\gamma\tau^*,\gamma\tau'') = \delta.$$

So by the triangle inequality,

$$\underline{\tau}'' \ge \tau^* - \delta.$$

Assume τ'' is chosen so that δ is small enough to satisfy

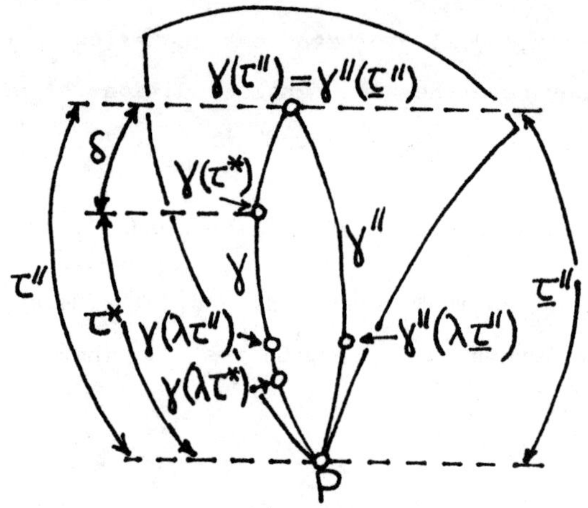

Figure 7

(3.17.1) $\delta < (\tau^* - \delta)\underline{r} \leq \underline{\tau}''\underline{r}$,

where $\underline{r}$ is as in (3.13.1). Applying (3.13.2) to γ'' and $\gamma\restriction[0,\tau^*]$ gives that for every $\lambda \in (0,1]$,

$$(3.17.2)\quad d^M(\gamma(\lambda\tau^*),\gamma''(\lambda\underline{\tau}'')) < 2\lambda H(\delta + \underline{\tau}''\underline{r})/h$$
$$< \lambda\underline{\tau}''\underline{r}^*/2,$$

by (3.17.1) and (3.13.1). Now

$$(3.17.3)\quad d^M(\gamma(\lambda\tau''),\gamma(\lambda\tau^*)) \leq \lambda(\tau'' - \tau^*)$$
$$= \lambda\delta$$
$$< \lambda\underline{\tau}''\underline{r}, \qquad \text{by (3.17.1)}$$
$$< \lambda\underline{\tau}''\underline{r}^*/4,$$

by (3.13.1). Hence (3.17.2) and (3.17.3) give

$$d^M(\gamma(\lambda\tau''),\gamma''(\lambda\underline{\tau}'')) < 3\lambda\underline{\tau}''\underline{r}^*/4.$$

It follows from Proposition 3.12 that $\gamma\restriction[0,\tau''] = \gamma''\restriction[0,\underline{\tau}'']$. That is, $\gamma\restriction[0,\tau'']$ is a geodesic, contradicting the definition of τ^*. Therefore $\tau^* = \underline{\tau}^{\#}$, and γ has been extended to $[0,\underline{\tau}^{\#}]$. q.e.d.

CHAPTER 4: e_P^{-1} IS INFINITESIMALLY ONE-TO-ONE

Introduction

The first consequences of hypothesis VII of (1.6.3) are studied in this chapter, the main result of which is that e_P^{-1} is "infinitesimally one-to-one" in a sense to be described. (1.6.3,VII) expresses, in terms of a given chart, a lower estimate on the sectional curvature K^M (Corollary 4.1.4). The intent is to use Rauch's comparison theorem, but first I need a model space $\mathcal{M}$ and a point $\mathcal{P}$ with which to compare M at P. Such a model is constructed in (4.4) and is shown to have an isolated conical singularity at $\mathcal{P}$ (Proposition 4.9). Its sectional curvature $K^{\mathcal{M}}$ is calculable (Lemma 4.8) and its geodesics are easy to describe (Lemma 4.10). Now K^M is squeezed between $K^{\mathcal{M}}$ and $\mathcal{K}$ (of (1.6.3,VI)), the sectional curvature of an n-sphere $\mathcal{S}$ of radius $\mathcal{K}^{-\frac{1}{2}}$. A boundary-value form of the comparison theorem is applied in the following context. Let γ be a geodesic in M from P with domain [0,1] and let $J \in T_{\gamma 1}M$ be fixed. For each $\nu \in (0,1)$ let $\underline{J}_\nu$ be the Jacobi field along $\gamma \restriction [\nu,1]$ such that $\underline{J}_\nu(\nu) = 0$, $\underline{J}_\nu(1) = J$. (Here and subsequently I abbreviate $\underline{J}(\gamma\mu)$ to $\underline{J}(\mu)$.) Comparing M to $\mathcal{S}$ shows that the vector field $\underline{J} = \lim_{\nu \to 0^+} \underline{J}_\nu$ is defined along $\gamma \restriction (0,1]$ (Theorem 4.13); and that $|\underline{J}(\mu)| \leq (3/2)\mu|J|$ (Corollary 4.15.8). Moreover $\underline{J}$ is a Jacobi field along γ with final value J at $\gamma(1)$ and "initial value 0" at $\gamma(0) = P$ in the sense that $\lim_{\mu \to 0^+} |\underline{J}(\mu)| = 0$ (Theorem 4.13). The statement that e_P^{-1} is "infinitesimally one-to-one" can be partly explained as saying that the "derivative" of $\underline{J}$ at P, "$(D_\gamma^M, \underline{J})(0)$," is non-zero provided that $J \neq 0$. To make sense, if not of this derivative, at least of its norm, I want to make use of $\lim_{\mu \to 0^+} |\underline{J}(\mu)|/\mu$; but to prove that this limit exists requires comparing M to $\mathcal{M}$. In $\mathcal{M}$, Jacobi fields from $\mathcal{P}$ along geodesics from $\mathcal{P}$ can be calculated explicitly (Lemma 4.18). The analogue of Corollary 4.15.8 is: $|\underline{J}(\mu)| \geq (1/2)\mu|J|$ (see Corollary 4.19.5). A more careful calculation shows that $|\underline{J}|_P$, defined to be $\lim_{\mu\to 0^+} |\underline{J}(\mu)|/\mu$, exists and is bounded between $(1/2)|J|$ and $(3/2)|J|$ (Proposition 4.21). In saying that e_P^{-1} is "infinitesimally one-to-one," this is the result to which I have been alluding, but rather prematurely so. For it still has to be shown that a Jacobi field along γ such as $\underline{J}$ is indeed the infinitesimal form of a one-parameter variation of γ through geodesics from P. This will be done in Chapter 5. Proposition 4.21 also shows that $||_P$ is a norm on the vector space $T_{\gamma 1}M$. This prepares for results of Chapter 6, in which $T_{\gamma 1}M$ will be identified with $T_{\hat{\gamma}}(T_PM)$ via $\exp_P$. Hence a norm $||_P$ can be defined on $T_{\hat{X}}(T_PM)$ for every $\hat{X} \in T_PM$, so that $(T_PM)^o$ has at least the rudiments of a Riemannian metric.

It will be generally assumed henceforth that P is an isolated conical singularity of M; that is, that (1.6.1), (1.6.2) and I -- VII of (1.6.3) hold. However, certain results -- those derived from comparing M to S -- do not depend on (1.6.3,VII), as will be mentioned in their statements.

Curvature Estimates

LEMMA 4.1. Let P be an isolated conical singularity of M. Let $X \in B^o(P,M;\underline{\tau}^{\#})$, where $\underline{\tau}^{\#}$ is as in Theorem 3.3. Let $\tau(X)$ be as in (2.15.4), let $\underline{\partial}_\tau$ be the unit radial vector field of (3.4.4) and let R^M be the curvature tensor on M^o (see (1.15.4)). Then for any $\psi,\omega \in T_XM$,

(1) $\|R^M(\psi,\omega)\| = |\psi\vee\omega|O(\tau(X)^{-2})$;

(2) $\|R^M(\underline{\partial}_\tau,\omega)\| = |\omega|O(\tau(X)^{-2+k})$;

(3) $|R^M(\underline{\partial}_\tau,\omega)\underline{\partial}_\tau| = |\omega|O(\tau(X)^{-2+2k})$.

Here k is as in V and VII of (1.6.3).

Proof of (1). This follows from (1.6.3,VII(i)) and Lemma 2.18.1.

Proof of (2). As in (2.19), write

$$\underline{\partial}_\tau(X) = \phi(X)\underline{\rho}^*(X) + \underline{X}(X),$$

where $\underline{\rho}$ is as in (2.15.3), $\phi = \langle \underline{\partial}_\tau,\underline{\rho}^*\rangle$ and $\underline{X} = \underline{\rho}^{*\perp}(\underline{\partial}_\tau)$. Then

$$R^M(\underline{\partial}_\tau,\omega) = \phi R^M(\underline{\rho}^*,\omega) + R^M(\underline{X},\omega).$$

Hence, since $|\phi| \leq 1$,

$$\|R^M(\underline{\partial}_\tau,\omega)\| \leq \|R^M(\underline{\rho}^*,\omega)\| + \|R^M(\underline{X},\omega)\|$$
$$\leq |\underline{\rho}|^{-1}\|R^M(\underline{\rho},\omega)\| + |\underline{X}\gamma\omega|O(\tau(X)^{-2}), \qquad \text{by (1)}$$
$$\leq |\underline{\rho}|^{-1}|\omega|O(\tau(X)^{-2+k}) + |\underline{X}||\omega|O(\tau(X)^{-2}),$$

by (1.6.3,VII(ii)) and Lemma 2.18.1. Now by (3.4.4), $\underline{\partial}_\tau(X) = \gamma^{X\prime}(\tau(X))$; and so by Lemma 2.20.2 applied to γ^X,

(4.1.4) $|\underline{X}(X)| = O(\tau(X)^k)$.

Also, by Corollary 2.18.8,

(4.1.5) $|\underline{\rho}(X)|^{-1} = 1 + O(\tau(X)^{2k})$.

Assertion (2) now follows.

Proof of (3).

$$R^M(\underline{\partial}_\tau,\omega)\underline{\partial}_\tau = \phi^2R^M(\underline{\rho}^*,\omega)\underline{\rho}^* + \phi R^M(\underline{\rho}^*,\omega)\underline{X} + \phi R^M(\underline{X},\omega)\underline{\rho}^* + R^M(\underline{X},\omega)\underline{X}.$$

By the Bianchi identity,

$$R^M(\underline{X},\omega)\underline{\rho}^* = R^M(\underline{\rho}^*,\omega)\underline{X} - R^M(\underline{\rho}^*,\underline{X})\omega.$$

Hence

$$R^M(\underline{\partial}_\tau,\omega)\underline{\partial}_\tau = \phi^2R^M(\underline{\rho}^*,\omega)\underline{\rho}^* + 2\phi R^M(\underline{\rho}^*,\omega)\underline{X} + \phi R^M(\underline{\rho}^*,\underline{X})\omega + R^M(\underline{X},\omega)\underline{X}.$$

Since $|\phi| \leq 1$,

$$|R^M(\underline{\partial}_\tau,\omega)\underline{\partial}_\tau| \leq |\underline{\rho}|^{-2}|R^M(\underline{\rho},\omega)\underline{\rho}| + 2|\underline{X}||\underline{\rho}|^{-1}\|R^M(\underline{\rho},\omega)\| + |\omega||\underline{\rho}|^{-1}\|R^M(\underline{\rho},\omega)\| + |\underline{X}|\|R^M(\underline{X},\omega)\|.$$

By (1.6.3,VII(iii)) and Lemma 2.18.1,

$$|R^M(\underline{\rho},\omega)\underline{\rho}| = |\omega|O(\tau(X)^{-2+2k}).$$

Using this, (1) and (2), and the estimates (4.1.4) and (4.1.5), assertion (3) now follows. q.e.d.

COROLLARY 4.1.6. Let the hypotheses be as in Lemma 4.1, and let Π be a tangent 2-plane at X. Then:

(1) $K^M(\Pi) = O(\tau(X)^{-2})$;

(2) if $\underline{\partial}_\tau(X) \in \Pi$, then $K^M(\Pi) = O(\tau(X)^{-2+k})$.

Here K^M denotes sectional curvature, as in (1.15.6).

<u>Proof of (1)</u>. Let ψ,ω be an orthonormal basis for Π. Then

$$\begin{aligned} |K^M(\Pi)| &= |\langle R^M(\psi,\omega)\psi,\omega\rangle|, \quad \text{by (1.15.6)} \\ &\leq \|R^M(\psi,\omega)\|\,|\psi|\,|\omega| \\ &= O(\tau(X)^{-2}), \end{aligned}$$

by Lemma 4.1.1.

<u>Proof of (2)</u>. Let $\underline{\partial}_\tau(X),\omega$ be an orthonormal basis for Π. Then

$$\begin{aligned} |K^M(\Pi)| &= |\langle R^M(\underline{\partial}_\tau,\omega)\underline{\partial}_\tau,\omega\rangle| \\ &\leq |R^M(\underline{\partial}_\tau,\omega)\underline{\partial}_\tau|\,|\omega| \\ &= O(\tau(X)^{-2+2k}), \end{aligned}$$

by Lemma 4.1.3. q.e.d.

<u>A Class of Models M</u>

(4.2) In (4.4) will be constructed a class of spaces M with isolated conical singularity P, one for each set of choices of:

(4.2.1) a closed, compact (n-1)-manifold L (as in (1.6.2)) with a Riemannian metric satisfying (3.2.1);

(4.2.2) an integer $q \geq 2$; and

(4.2.3) a positive number A.

In Proposition 4.19 a given M with isolated conical singularity P will be compared to any M chosen subject to the following constraints:

(4.2.4) L is the base of every chart at P (see Corollary 2.21.4), with arbitrary Riemannian metric;

(4.2.5) $q \geq 1/2k$, k being as in V and VII of (1.6.3);

and whenever $X \in B^{o}(P.M;\underline{\tau}^{\#})$ and Π is a tangent 2-plane at X containing the vector $\underline{\partial}_{\tau}(X)$ (see (3.4)), then

$$(4.2.6)\quad K^{M}(\Pi) > -A(\tau X)^{-2+2k} \geq -A(\tau X)^{-2+1/q},$$

since $\tau(X) < \underline{\tau}^{\#} < 1$ by Theorem 3.3. Corollary 4.1.4 assures us that A can be chosen to satisfy (4.2.6).

Whenever L, q and A are chosen as in (4.2.1) -- (4.2.3), a corresponding M is constructed by means of a function $r(\tau)$ determined by the o.d.e.

$$(4.2.7)\quad r''(\tau) = A\tau^{-2+1/q}r(\tau);\ \lim_{\tau\to 0^{+}} r(\tau) = 0;\ \lim_{\tau\to 0^{+}} r'(\tau) = 1.$$

LEMMA 4.3. Given q and A as in (4.2.2) and (4.2.3), then $\underline{\tau}^{\#} \in (0,1)$ can be chosen so small that (4.2.7) has a solution on $(0,\underline{\tau}^{\#}]$ of the form

(1) $r(\tau) = \tau(1 + \sum_{1}^{\infty} a_m\tau^{m/q})$ for certain constants a_m, and satisfying:

(2) $r'(\tau) \in [4/5,6/5]$ on $(0,\underline{\tau}^{\#}]$;

(3) $(4/5)\tau \leq r(\tau) \leq (6/5)\tau$ on $(0,\underline{\tau}^{\#}]$.

Proof. The substitution $\tau = t^{q}$ transforms (4.2.7) into

$$(4.3.4)\quad r''(t) = (q-1)t^{-1}r'(t) + q^{2}At^{-1}r(t),$$

which has a regular singularity at $t = 0$. The indicial equation is

$$\nu(\nu-1) = (q-1)\nu,$$

with roots $\nu = q$ and $\nu = 0$. Hence (see Hartman [7, p. 85]) there is a solution of (4.3.4) of the form

(4.3.5) $r(t) = t^q\left(1 + \sum_1^\infty a_m t^m\right),$

convergent on some $[0,\varepsilon]$ with $\varepsilon > 0$. Choose $\underline{\tau}^\# < \varepsilon^q$; then substituting $t = \tau^{1/q}$ in (4.3.5) gives (1). Condition (2) can be met by choosing $\underline{\tau}^\#$ small enough; and (3) follows from (2) and (4.2.7). q.e.d.

(4.4) Construction of M. Let L, q and A be as in (4.2.1) -- (4.2.3) and let r be as in Lemma 4.3. By (4.2.7) and (3.2.5),

$$K^L < 1 = \lim_{\tau\to 0^+} r'(\tau)^2;$$

hence $\underline{\tau}^\#$ can be chosen so small that

(4.4.1) $K^L < \inf_{(0,\underline{\tau}^\#]} r'(\tau)^2.$

Let M be the cone $c^{\underline{\tau}^\#}L$, set $P = c$ and give M^o the metric

(4.4.2) $(ds^M)^2(Z,t) = r(\tau)^2(ds^L)^2(Z) + d\tau^2.$

(4.5) Local coordinates on M^o. Fix $(Z^*,\tau^*) \in M^o$, let $x_1,\ldots,x_{n-1}$ be normal coordinates near Z^* in L centered at Z^* (see Hicks [8, p. 133]), and set $x_n = \tau-\tau^*$. Then, in the notation of (1.15),

(4.5.1) $g^L_{ij}(Z^*,\tau^*) = g^L_{ij}(0,\ldots,0) = \delta^j_i$ for $i,j = 1,\ldots,n-1$;

(4.5.2) $\Gamma^{Lj}_{ik}(Z^*,\tau^*) = 0$ for $i,j,k = 1,\ldots,n-1$.

Further

(4.5.3)
$$\left.\begin{aligned} g^M_{ij}(x_1,\ldots,x_n) &= r(\tau^*+x_n)^2 g^L_{ij}(x_1,\ldots,x_{n-1}),\\ g^M_{in} &\equiv 0, \end{aligned}\right\}\text{ for } i,j \le n-1$$
$$g^M_{nn} \equiv 0.$$

In particular,

$$(4.5.4)\quad g^M_{ij}(Z^*,\tau^*)\ \begin{cases} r(\tau^*)^2\delta^j_i, & \text{for } i,j \le n-1 \\ \delta^j_n, & \text{for } i = n. \end{cases}$$

LEMMA 4.6. In the special coordinate system of (4.5), and with $i,j,k \le n-1$,

(1) $\Gamma^{Mj}_{ik} = \Gamma^{Lj}_{ik}$;

(2) $\Gamma^{Mj}_{nk} = \Gamma^{Mj}_{kn} = (r'/r)\delta^j_k$;

(3) $\Gamma^{Mn}_{ik} = -rr'g^L_{ik}$;

(4) all other components of Γ^M are identically 0.

Proof. The formula for Γ^M in terms of g^M is given, for example, in Hicks [8, p. 72]. The proof is now a straightforward calculation, which I omit.

q.e.d.

LEMMA 4.7. In the special coordinate system of (4.5), and with $h,i,j,k \le n-1$, the curvature tensor R^M_{hkij}, evaluated at (Z^*,τ^*), has the following form:

(1) $R^M_{hkij}(Z^*,\tau^*) = r^2R^L_{hkij}(Z^*) + r^2r'^2(\delta^i_h\delta^j_k - \delta^i_k\delta^j_n)$.

(2) If in (1) any one index is replaced by n, then the result is 0; e.g.

$$R^M_{hkin}(Z^*,\tau^*) = 0.$$

(3) $R^M_{nknj}(Z^*,\tau^*) = rr''\delta^j_k$, and a similar result (up to $\pm$) holds whenever one of h and k and one of i and j are replaced by n.

(4) All other components of R^M are identically 0.

Proof of (1). From (1.15.4) (or Hicks [8, p. 63]),

$$(4.7.5)\quad R^M_{hkij} = \sum_{a=1}^{n}\Big\{\underline{\partial}_h\Gamma^{Ma}_{ik} - \underline{\partial}_k\Gamma^{Ma}_{ih}$$

$$+ \sum_{b=1}^{n}(\Gamma^{Mb}_{ik}\Gamma^{Ma}_{bh} - \Gamma^{Mb}_{ih}\Gamma^{Ma}_{bk})\Big\}g^M_{aj}.$$

By (4.5.3) the summand with $a = n$ gives 0. Using Lemma 4.6.1 it follows that the terms with $a,b = 1,\ldots,n-1$ give $r^2R^L_{hkij}$. The remaining terms are:

$$\sum_{a=1}^{n}(\Gamma^{Mn}_{ik}\Gamma^{Ma}_{nh} - \Gamma^{Mn}_{ih}\Gamma^{Ma}_{nk})g^M_{am}.$$

Using (4.5.1), (4.5.4) and Lemma 4.6.2 and 4.6.3, this reduces, at (Z^*,τ^*), to the second term on the right-hand side of (1).

Proof of (2). In view of the symmetries of R (combine (1.15.5) with Hicks [8, p. 72]), it is sufficient to prove (2) in the case cited. When $j = n$ in (4.7.5) only the sum and with $a = n$ is non-zero. When $a = n$, the term in $\sum_b$ with $b = n$ is 0 because $\Gamma^{Mn}_{nh} = \Gamma^{Mn}_{nk} = 0$ by Lemma 4.6.4. The terms in $\sum_b$ with $b \leq n-1$ are 0 at (Z^*,τ^*) by Lemma 4.6.1 and (4.5.2). Thus

$$(4.7.6)\quad R^M_{hkin}(Z^*,\tau^*) = (\underline{\partial}_h\Gamma^{Mn}_{ik} - \underline{\partial}_k\Gamma^{Mn}_{ih})(Z^*,\tau^*)$$

$$- rr'(\underline{\partial}_h g^L_{ik} - \underline{\partial}_k g^L_{ih})(Z^*,\tau^*).$$

Now

$$\underline{\partial}_h g^L_{ik} = \langle D^L_{\underline{\partial}_h}\underline{\partial}_i,\underline{\partial}_k\rangle + \langle\underline{\partial}_i,D^L_{\underline{\partial}_h}\underline{\partial}_k\rangle \quad \text{(see GKM [6, p. 79])}$$

$$= \sum_{c=1}^{n-1}(\Gamma^{Lc}_{ih}g^L_{ck} + \Gamma^{Lc}_{ik}g^L_{ch}),$$

which reduces at (Z^*,τ^*) to $(\Gamma^{Lk}_{ih} + \Gamma^{Lh}_{ik})(Z^*,\tau^*)$, which is 0 by (4.5.2). That is,

$$\underline{\partial}_h g^L_{ik}(Z^*,\tau^*) = 0.$$

Similarly

$$\underline{\partial}_k g^L_{ih}(Z^*,\tau^*) = 0;$$

and putting these into (4.7.6) proves (2).

Proof of (3). This is done by a straight-forward calculation.

Proof of (4). Under the hypothesis of (4), either $h = k = n$ or $i = j = n$; in either case the component of R^M is identically 0 by the symmetry properties of the curvature tensor. q.e.d.

LEMMA 4.8. With the special coordinate system of (4.5), let $\Pi^L(j,i)$ be the 2-plane at Z^* spanned by $\underline{\partial}_i$ and $\underline{\partial}_j$, and let $\Pi^M(h,k)$ be the 2-plane at (Z^*,τ^*) spanned by $\underline{\partial}_k$ and $\underline{\partial}_h$; here $i,j = 1,\ldots,n-1$; $h,k = 1,\ldots,n$; and $i \neq j$, $h \neq k$. Then

(1) $K^M(\Pi^M(j,i))(Z^*,\tau^*) = \{K^L(\Pi^L(j,i))(Z^*) - r'(\tau^*)^2\}/r(\tau^*)^2$;

(2) $K^M(\Pi^M(n,i))(Z^*,\tau^*) = -A(\tau^*)^{-2+1/q}$.

Proof of (1). Since the vectors $\{\underline{\partial}_h(Z^*,\tau^*)\}$ are mutually orthogonal,

$$K^M(\Pi^M(h,k))(Z^*,\tau^*) = (R^M_{hkkh}/|\underline{\partial}_k|^2|\underline{\partial}_h|^2)(Z^*,\tau^*). \tag{4.8.3}$$

By (4.5.4),

$$K^M(\Pi^M(j,i))(Z^*,\tau^*) = R^M_{jiij}(Z^*,\tau^*)/r(\tau^*)^4.$$

The vectors $\{\underline{\partial}_i(Z^*)\}$ form an orthonormal basis of $T_{Z^*}L$. Hence

$$K^L(\Pi^L(j,i))(Z^*) = R^L_{jiij}(Z^*).$$

Now (1) follows from Lemma 4.7.1.

Proof of (2). When $h = n$ and $k = i$ in (4.8.3),

$$|\underline{\partial}_i(Z^*,\tau^*)| = r(\tau^*) \text{ and } |\underline{\partial}_n(Z^*,\tau^*)| = 1.$$

Hence

$$K^M(\Pi^M(n,i))(Z^*,\tau^*) = R^M_{niin}(Z^*,\tau^*)/r(\tau^*)^2$$

$$= -R_{nini}(Z^*,\tau^*)/r(\tau^*)^2$$

$$= -r''(\tau^*)/r(\tau^*), \quad \text{by Lemma 4.7.3}$$

$$= -A(\tau^*)^{-2+1/q}, \quad \text{by (4.2.7)}$$

q.e.d.

Proposition 4.9. P is an isolated conical singularity of M. Moreover one may take $k = 1/2q$ in (1.6.3,V and VII), and any $K > 0$ in (1.6.3,VI).

Proof. Clearly (1.6.1) holds. Let $h\colon c^{\tau}_{-}L^{\#} \to M$ be the identity map; then (1.6.2) holds, and it remains to verify I -- VII of (1.6.3). Hypotheses I -- IV are easy to check, since $Dh\cdot\underline{\partial}_t = \underline{\partial}_\tau$. In the rest of the proof I use the terminology of (4.5).

Proof of V. It suffices to prove V when $\psi \in T_{(Z^*,\tau^*)}M^o$ is either $\underline{\partial} = \underline{\partial}_n(Z^*,\tau^*)$ or a unit vector orthogonal to $\underline{\partial}_n(Z^*,\tau^*)$. In the former case,

$$D^M_\psi(Dh\cdot\underline{\partial}_t) = D^M_{\underline{\partial}_n}\underline{\partial}_n(Z^*,\tau^*)$$

$$= \sum_{j=1}^{n} \Gamma^{Mj}_{nn}\underline{\partial}_j(Z^*,\tau^*)$$

$$= 0,$$

by Lemma 4.6.4; and also

$$(Dh\cdot\underline{\partial}_t)^{\perp}(\psi) = \underline{\partial}_n^{\perp}(\underline{\partial}_n) = 0;$$

so V holds in this case. In the latter case, the special coordinate system of (4.5) can be chosen so that $\psi = \underline{\partial}_1(Z^*,\tau^*)$. Then

$$D^M_\psi(Dh\cdot\underline{\partial}_t) = D^M_{\underline{\partial}_1}\underline{\partial}_n(Z^*,\tau^*)$$

$$= \sum_{j=1}^{n} \Gamma^{Mj}_{n1}\underline{\partial}_j(Z^*,\tau^*)$$

$$= (r'/r)(\tau^*)\underline{\partial}_1(Z^*,\tau^*),$$

by Lemma 4.6.2. By Lemma 4.3,

$$(r'/r)(\tau^*) = (\tau^*)^{-1}(1 + O(\tau^{*1/q})).$$

Hence

$$D_{\psi}^{M}(Dh\cdot\underline{\partial}_{t})(Z^*,\tau^*) = (\tau^*)^{-1}\psi + O(\tau^{*-1+1/q})\psi.$$

Since $(Dh\cdot\underline{\partial}_{t})^{\perp}(\psi) = \psi$, V holds in this case too.

Proof of VI. In the notation of Lemma 4.8, (4.4.1) implies that for $h,k = 1,\ldots,n$,

$$K^{M}(\Pi^{M}(h,k))(Z^*,\tau^*) \le 0$$

for every tangent coordinate 2-plane at (Z^*,τ^*). Therefore $K^{M}(\Pi) < 0$ for every tangent 2-plane at (Z^*,τ^*). This proves VI with any $K > 0$.

Proof of VII(i). Since the metric on L is of class C^{∞}, its curvature tensor R^{L} is bounded above in norm. That is, there exists R such that whenever $Z \in L$ and $v,w \in T_{Z}L$, then

$$\| R^{L}(v,w)\| \le R|v\vee w|.$$

In particular, whenever $h,i,j,k \le n-1$,

$$|R^{L}_{hkij}(Z^*)| \le \| R^{L}(\underline{\partial}_{h},\underline{\partial}_{k})\|(Z^*) \le R.$$

It follows from Lemma 4.7 that when $h,k \le n-1$,

$$(4.9.1)\quad |R^{M}_{hkij}(Z^*,\tau^*)| \begin{cases} = 0 & \text{unless } i,j \le n-1 \\ \le r(\tau^*)^{2}(R + r'(\tau^*)^{2}O(\tau^{*0})), & \text{when } i,j \le n-1 \end{cases}$$
$$= r(\tau^*)^{2}O(\tau^{*0})).$$

(For $O(\tau^{*0})$ see (1.12.1).) When $i,j \le n-1$, $\underline{\partial}_{i}(Z^*,\tau^*)$ and $\underline{\partial}_{j}(Z^*,\tau^*)$ are orthogonal vectors of length $r(\tau^*)$, by (4.5.4). So

$$(4.9.2)\quad |\underline{\partial}_{i}\vee\underline{\partial}_{j}|(Z^*,\tau^*) = r(\tau^*)^{2}, \quad \text{for } i,j \le n-1.$$

Hence, by (4.9.1), for $h,k \le n-1$,

$$\| R^{M}(\underline{\partial}_{h},\underline{\partial}_{k})\|(Z^*,\tau^*) \le \sup_{i,j\le n-1} (|R^{M}_{hkij}|/|\underline{\partial}_{i}\vee\underline{\partial}_{j}|)(Z^*,\tau^*)$$

$$= O(\tau^{*0})$$

$$= r(\tau^*)^{-2}O(\tau^{*0})\,|\underline{\partial}_h \vee \underline{\partial}_k|, \quad \text{by } (4.9.2)$$

$$= O(\tau^{*-2})\,|\underline{\partial}_h \vee \underline{\partial}_k|,$$

by Lemma 4.3.1. Once VII(ii) is proved, we shall have that for all $h,k = 1,\ldots,n,$

$$\| R^M(\underline{\partial}_h,\underline{\partial}_k) \| (Z^*,\tau^*) = O(\tau^{*-2})\,|\underline{\partial}_h \vee \underline{\partial}_k|,$$

which implies VII(i). Thus VII(i) has been reduced to VII(ii).

Proof of VII(ii) and (iii). From Lemma 4.7,

$$R^M_{nkij}(Z^*,\tau^*) = \begin{cases} (rr'')(\tau^*), & \text{if } (i,j) = (n,k) \text{ and } k \leq n-1 \\ -(rr'')(\tau^*), & \text{if } (i,j) = (k,n) \text{ and } k \leq n-1, \\ 0, & \text{for other } i,j \text{ and } k. \end{cases}$$

Now $\underline{\partial}_n(Z^*,\tau^*)$ is a unit vector and, for $k \leq n-1$, $\underline{\partial}_k(Z^*,\tau^*)$ is orthogonal to it and has length $r(\tau^*)$. Therefore

$$|\underline{\partial}_n \vee \underline{\partial}_k|(Z^*,\tau^*) = |\underline{\partial}_k(Z^*,\tau^*)| = r(\tau^*).$$

It follows that, for $k \leq n-1$,

$$\begin{aligned} \| R^M(\underline{\partial}_n,\underline{\partial}_k) \| (Z^*,\tau^*) &= |R^M(\underline{\partial}_n,\underline{\partial}_k)\underline{\partial}_n|(Z^*,\tau^*) \\ &= (rr'')(\tau^*)/|\underline{\partial}_k(Z^*,\tau^*)| \\ &= r''(\tau^*) \\ &= (r''/r)(\tau^*)\,|\underline{\partial}_k(Z^*,\tau^*)| \\ &= O(\tau^{*-2+1/q})\,|\underline{\partial}_k(Z^*,\tau^*)|, \end{aligned}$$

by (4.2.7). Since $R^M(\underline{\partial}_n,\underline{\partial}_n) \equiv 0$, it follows that VII(ii) and (iii) hold, with $k = 1/q$ and $k = 1/2q$ respectively, whenever ω is a coordinate tangent vector at (Z^*,τ^*). Therefore they both hold, with $k = 1/2q$, for general $\omega \in T_{(Z^*,\tau^*)}M$. q.e.d

LEMMA 4.10. The geodesics from P in M are precisely the curves $g(\lambda) = (Z^*,\lambda\tau^*)$ for $(Z^*,\tau^*) \in c^\infty L$ and $\lambda \in [0,\underline{\tau}^\#/\tau^*)$.

Proof. Let $Z^* \in L$ and set $g(\tau) = (Z^*,\tau)$ for $\tau \in [0,\tau^*]$, where $\tau^* \in (0,\underline{\tau}^\#)$ is fixed. Let $g^*(\sigma) = (Z\sigma,\tau\sigma)$ be a geodesic from P to (Z^*,τ^*) parametrized by arc length with domain $[0,\sigma^*]$. Then g^* is C^∞ by Proposition 2.13, so $\tau(\sigma)$ is differentiable. Now

$$\begin{aligned}\tau'(\sigma) &= \langle g^{*\prime}(\sigma), \underline{\partial}_\tau \rangle \\ &= |\underline{\partial}_\tau{}^{\parallel}(g^{*\prime}\sigma)| \\ &\le |g^{*\prime}(\sigma)| \\ &= 1,\end{aligned}$$

with equality only if $g^{*\prime}(\sigma) = \underline{\partial}_\tau{}^{\parallel}(g^*\sigma)$. Hence

$$\begin{aligned}\tau^* &= \int_0^{\sigma^*} \tau'(\sigma)\,d\sigma \\ &\le \int_0^{\sigma^*} |g^{*\prime}(\sigma)|\,d\sigma \\ &= L(g^*) \\ &= \sigma^*;\end{aligned}$$

and since $g^{*\prime}$ is continuous, equality holds only if

$$g^{*\prime}(\sigma) = \underline{\partial}_\tau\,(g^*\sigma) \quad \text{for all} \quad \sigma \in (0,\sigma^*].$$

Therefore $g^* = g$. This shows that the geodesics from P in M parametrized by arc length are precisely the paths $g(\tau) = (Z^*,\tau)$ on $[0,\underline{\tau}^\#)$, for $Z^* \in L$. The lemma now follows. q.e.d.

(4.10.1) Remark. It would not be difficult to prove directly that the identity map

$$h\colon\ c^{\underline{\tau}^\#} L \to M,$$

used as a chart at P in the proof of Proposition 4.9, is actually the exponential map of M at P, satisfying (1.1) -- (1.5). This program would not use (1.6.3,VI). In fact one could construct M using any L satisfying (4.2.1) and any $A < 0$ in (4.2.3); then by Lemma 4.9.2, (1.6.3,VI) would definitely be violated, yet (1.1) -- (1.5) would still hold. This example extends Example 3.2, which corresponds to taking $A = 0$ in (4.2.3), and again suggests that (1.6.3,VI) can be weakened.

Jacobi Fields

(4.11) Notation. Let γ be a weak geodesic in M such that $\gamma\colon (0,1] \to M^o$. The symbol $\underline{J}$ will denote a Jacobi field along $\gamma \restriction (0,1]$ or along γ, according as $\gamma(0)$ equals P or not. Thus

(4.11.1) $\underline{J}(\mu) \in T_{\gamma\mu}M$, and

(4.11.2) $(D^M_{\gamma'})^2\underline{J} = R^M(\gamma',\underline{J})\gamma'$.

Let $\nu \in (0,1]$ or $\nu \in [0,1]$, according as $\gamma(0)$ equals P or not. $\underline{J}_\nu$ will denote a Jacobi field along $\gamma \restriction [\nu,1]$ such that

(4.11.3) $\underline{J}_\nu(\nu) = 0$.

$\underline{J}$ will be called a Jacobi field along γ from P if

(4.11.4) $\lim_{\mu\to 0^+} |\underline{J}(\mu)| = 0$, or $|\underline{J}(0)| = 0$,

according as $\gamma(0)$ equals P or not.

In the rest of (4.11), but not subsequently, I shall write $\underline{J}_0$ for $\underline{J}$, thus allowing $\nu = 0$ and subsuming (4.11.4) under (4.11.3). Since γ is a weak geodesic, $D^M_{\gamma'}\gamma' \equiv 0$. It follows that, for $\mu \in [\nu,1]$,

$$\frac{d}{d\mu}\langle \underline{J}_\nu(\mu),\gamma'\mu\rangle = \langle D^M_{\gamma'\mu}\underline{J}_\nu,\gamma'\mu\rangle,$$

and similarly that

(4.11.5) $$\frac{d^2}{d\mu^2}\langle \underline{J}_\nu(\mu),\gamma'\mu\rangle = \langle (D^M_{\gamma'})^2\underline{J}_\nu,\gamma'\mu\rangle$$
$$= \langle R^M(\gamma',\underline{J}_\nu)\gamma',\gamma'\mu\rangle, \quad \text{by (4.11.2)}$$
$$= 0.$$

From (4.11.5) and (4.11.3) it follows that if, for some $\underline{\lambda} \in (\nu,1]$, $\underline{J}_\nu(\underline{\lambda})$ is orthogonal to $\gamma'(\underline{\lambda})$, then

(4.11.6) $\langle \underline{J}_\nu(\mu), \gamma'\mu \rangle \equiv 0.$

A Jacobi field satisfying (4.11.6) and (4.11.3) (or (4.11.4)) will be denoted ${}^{\perp}\underline{J}_\nu$.

Again by (4.11.5) and (4.11.3), if, for some $\underline{\lambda} \in (\nu,1]$, $\underline{J}_\nu(\underline{\lambda})$ is a scalar multiple $j(\underline{\lambda})\gamma'(\underline{\lambda})$ of $\gamma'(\underline{\lambda})$, then

$$\langle \underline{J}_\nu(\mu), \gamma'\mu \rangle = ((\mu-\nu)/(\underline{\lambda}-\nu))j(\underline{\lambda})|\gamma'\mu|^2.$$

In this case, let ${}^{\parallel}\underline{J}_\nu$ denote the Jacobi field

(4.11.7) ${}^{\parallel}\underline{J}_\nu(\mu) = ((\mu-\nu)/(\underline{\lambda}-\nu))j(\underline{\lambda})\gamma'(\mu).$

For general $\underline{J}_\nu$, let $\underline{\lambda} \in (\nu,1]$ be given, and set $J^{\underline{\lambda}} = \underline{J}_\nu(\underline{\lambda})$. Write

(4.11.8) $J^{\underline{\lambda}} = (\gamma'\underline{\lambda})^{\parallel}(J^{\underline{\lambda}}) + (\gamma'\underline{\lambda})^{\perp}(J^{\underline{\lambda}})$

in orthogonal components, as in (1.13.5). Let ${}^{\parallel}\underline{J}_\nu$ satisfy (4.11.7) with ${}^{\parallel}\underline{J}_\nu(\underline{\lambda}) = \gamma'(\underline{\lambda})^{\parallel}(J^{\underline{\lambda}})$. Then $\underline{J}_\nu - {}^{\parallel}\underline{J}_\nu$ is a Jacobi field along γ satisfying (4.11.3) and (4.11.6). Thus (4.11.8) may be extended to

(4.11.9) $\underline{J}_\nu = {}^{\parallel}\underline{J}_\nu + {}^{\perp}\underline{J}_\nu;$

and it is not hard to check that this decomposition does not depend on the choice of $\underline{\lambda}$. Conversely, if γ has no pair of points which are conjugate along γ -- as will turn out to be the case in the rest of this chapter, by Corollary 4.12.9 -- then (4.11.9) is the only decomposition of a given $\underline{J}_\nu$ into Jacobi fields along γ satisfying (4.11.3), such that one of them is a scalar multiple of γ' and the other is everywhere orthogonal to γ'.

(4.11.10) For the rest of this chapter, M will denote a space with isolated conical singularity P which satisfies (4.2) and (4.4); and S will denote an n-sphere of radius $K^{-\frac{1}{2}}$, where K is as in (1.6.3,VI).

T will denote a number such that

(4.11.11) $0 < T \leq \min\{\pi/(2K^{\frac{1}{2}}), \underline{\tau}^{\#}\},$

where $\underline{\tau}^{\#}$ is as in (4.4). Unless otherwise specified, I shall assume that γ

has domain [0,1], and that

(4.11.12) $L(\gamma) = |\gamma'| < T.$

Proposition 4.12. In addition to the notation of (4.11), assume that $\nu < \lambda \leq 1$, and set $\mathrm{J}^\lambda = \underline{\mathrm{J}}_\nu(\lambda)$. Then:

(1) For $\mu \in [\nu,\lambda]$,

$$\begin{aligned}|\underline{\mathrm{J}}_\nu(\mu)| &\leq ((\mu-\nu)/(\lambda-\nu))(1 + O((\lambda T)^2))|\mathrm{J}^\lambda| \\ &\leq (3/2)((\mu-\nu)/(\lambda-\nu))|\mathrm{J}^\lambda| \\ &\leq (3/2)(\mu/\lambda)|\mathrm{J}^\lambda|.\end{aligned}$$

(2) For $\mu \in [\lambda,1]$,

$$\begin{aligned}|\underline{\mathrm{J}}_\nu(\mu)| &\geq (\mu/\lambda)(1 + O(T^2))|\mathrm{J}^\lambda| \\ &\geq (1/2)(\mu/\lambda)|\mathrm{J}^\lambda|.\end{aligned}$$

(3) $|\mathrm{D}^M_{\gamma'}\underline{\mathrm{J}}_\nu|(\nu) \leq (3/2)(1/(\lambda-\nu))|\mathrm{J}^\lambda|.$

This result does not require (1.6.3,VII).

Proof of (1). By (4.11.9) and (4.11.7) it is enough to prove the assertion when $\underline{\mathrm{J}}_\nu = {}^{\perp}\underline{\mathrm{J}}_\nu$. Let g be a geodesic in S (of (4.11.10)) with domain [0,1] such that $|g'| = |\gamma'|$. Let $\underline{J}_\nu$ be a Jacobi field along $g\restriction[\nu,\lambda]$ such that (4.11.3) holds, $\underline{J}_\nu$ is orthogonal to g', and $|\underline{J}_\nu(\lambda)| = |\mathrm{J}^\lambda|$. Set $J^\lambda = \underline{J}_\nu(\lambda)$. Since $|g'| < T < \pi/(2K^{1/2})$ by (4.11.11) and (4.11.12), no two points of g are conjugate along g. In view of (1.6.3,VI), Rauch's comparison theorem applies. The main step in the proof of that theorem (see, for example, GKM [6, p. 178]) shows that $|\underline{\mathrm{J}}_\nu(\mu)|/|\underline{J}_\nu(\mu)|$ is non-decreasing in μ. This ratio is 1 when $\mu = \lambda$. Therefore

(4.12.4) $|\underline{\mathrm{J}}_\nu(\mu)| \leq |\underline{J}_\nu(\mu)|$, on $[\nu,\lambda]$.

Now

$$\begin{aligned}(4.12.5)\quad |\underline{J}_\nu(\mu)| &= |J^\lambda|\sin\{(\mu-\nu)|g'|K^{1/2}\}/\sin\{(\lambda-\nu)|g'|K^{1/2}\} \\ &= |J^\lambda|((\mu-\nu)/(\lambda-\nu))(1 + O(\{(\lambda-\nu)|g'|\}^2)).\end{aligned}$$

Since $(\lambda-\nu)|g'| < \lambda T$, the first estimate of (1) now follows from (4.12.4). To prove the succeeding estimates, note that $(\mu-\nu)|g'|K^{\frac{1}{2}}$ and $(\lambda-\nu)|g'|K^{\frac{1}{2}}$ are both $<\pi/2$, by (4.11.11) and (4.11.12); hence

$$
\begin{aligned}
(4.12.6)\qquad \sin\{(\mu-\nu)|g'|K^{\frac{1}{2}}\}/\sin\{(\lambda-\nu)|g'|K^{\frac{1}{2}}\} \\
&< (3/2)((\mu-\nu)/(\lambda-\nu)), \quad \text{by } (1.17.3) \\
&\le (3/2)(\mu/\lambda),
\end{aligned}
$$

by (1.17.4). The required estimates now follow from (4.12.5) and (4.12.4).

<u>Proof of (2)</u>. The proof is the same as that of (1), except that the inequality is reversed in (4.12.4) and (4.12.6) is replaced by (4.12.7), which again follows from (1.17.3) and (1.17.4):

$$(4.12.7)\qquad \sin\{(\mu-\nu)|g'|K^{\frac{1}{2}}\}/\sin\{(\lambda-\nu)|g'|K^{\frac{1}{2}}\} > (1/2)(\mu/\lambda).$$

<u>Proof of (3)</u>. Let e denote $\exp_{\gamma\nu}$. Set

$$\tilde{\gamma}(\mu) = (\mu-\nu)\gamma'(\nu);$$

so $e\circ\tilde{\gamma} = \gamma$. By (4.11.11), (4.11.12) and Proposition 3.7.2, De is nonsingular at each $\tilde{\gamma}(\mu)$, so I may define the vector field $\tilde{\underline{J}}_\nu$ along $\tilde{\gamma}$ by

$$\tilde{\underline{J}}_\nu(\mu) = (De)^{-1}(\gamma\mu)\cdot\underline{J}_\nu(\mu).$$

Since $\underline{J}_\nu$ is a Jacobi field and vanishes at $\gamma(\nu)$, it is the derivative of a one-parameter variation of γ through geodesics from $\gamma(\nu)$. Under e^{-1} these geodesics are transformed into rays from $\gamma(\nu)$ in $T_{\gamma\nu}M$. It follows that the vector field $\tilde{\underline{\Psi}}_\nu$, defined along $\gamma\restriction(\nu,1]$ by

$$\tilde{\underline{\Psi}}_\nu(\mu) = (\tilde{\underline{J}}_\nu(\mu))/(\mu-\nu),$$

is constant. Define $\tilde{\underline{\Psi}}_\nu(\nu)$ to have this same constant value; then the vector field $\underline{\Psi}_\nu = (De)\cdot\tilde{\underline{\Psi}}_\nu$ is parallel along $\gamma\restriction[\nu,1]$, and $\underline{J}_\nu(\mu) = (\mu-\nu)\underline{\Psi}_\nu(\mu)$. Hence

$$
\begin{aligned}
(4.12.8)\qquad (D^M_{\gamma'}\underline{J}_\nu)(\nu) &= \underline{\Psi}_\nu(\nu) \\
&= \lim_{\mu\to 0^+} (\underline{J}_\nu\mu)/(\mu-\nu).
\end{aligned}
$$

Assertion (3) now follows from the second inequality of (1). q.e.d.

COROLLARY 4.12.9. Let γ be a weak geodesic in M^o satisfying (4.11.12). Then no two points of γ are conjugate along γ. (1.6.3,VII) is not needed for this result.

COROLLARY 4.12.10. Let γ be a weak geodesic in M^o satisfying (4.11.12), and let $\underline{J}$ be a Jacobi field along $\gamma \upharpoonright [\nu,1]$ such that $\underline{J}(1) = 0$. Then:

(1) $|\underline{J}(\mu)| \leq (3/2)((1-\mu)/(1-\nu))|\underline{J}(\nu)|$, for $\mu \in [\nu,1]$;

(2) $|D^M_{\gamma'}\underline{J}|(1) \leq (3/2)|\underline{J}(\nu)|/(1-\nu)$.

(1.6.3,VII) is not needed for this result.

Proof. Define $\tilde{\gamma}$ on [0,1] by

$$\tilde{\gamma}(\lambda) = \gamma(1 - \lambda(1-\nu)),$$

so that

$$\gamma(\mu) = \tilde{\gamma}((1-\mu)/(1-\nu)).$$

Thus $\tilde{\gamma}$ is a reparametrization of γ as a weak geodesic from $\gamma(1)$ to $\gamma(\nu)$, and $\underline{J}$ may be regarded as a Jacobi field along $\tilde{\gamma}$ from $\tilde{\gamma}(0)$ in the sense of (4.11.4). Apply Proposition 4.12.1 to $\tilde{\gamma}$, with $\nu = 0$, λ replaced by 1 and μ by λ, to obtain

$$|\underline{J}|(\tilde{\gamma}\lambda) \leq (3/2)\lambda|\underline{J}|(\gamma 1).$$

Substituting $\lambda = (1-\mu)/(1-\nu)$ gives the first estimate. The second one is proved in the same way, using (4.12.8). q.e.d.

THEOREM 4.13. Let γ be a weak geodesic in M, as in (4.11), satisfying (4.11.12). Let $J \in T_{\gamma 1}M$ be fixed, and for each $\nu \in (0,1)$, let $\underline{J}_\nu$ be as in (4.11.3). Then:

(1) For each $\mu \in (0,1]$,

$$\underline{J}(\mu) = \lim_{\substack{\nu<\mu \\ \nu\to 0^+}} \underline{J}_\nu(\mu) \quad \text{exists in} \quad T_{\gamma\mu}M;$$

(2) The resulting vector field $\underline{J}$ along $\gamma\restriction(0,1]$ is a Jacobi field along γ from P with final condition J;

(3) $\underline{J}$ is the only Jacobi field along γ from P with final condition J.

This theorem does not require (1.6.3,VII).

(4.14) Notation. When the dependence of $\underline{J}$ on J is important, I shall write $\underline{J}\{J\}$.

(4.14.1) Let T_γ denote $\{\underline{J}\{J\}$ for $J \in \gamma(1)\}$. Theorem 4.13 implies that T_γ is a vector space, naturally isomorphic to $T_{\gamma 1}M$. It will be shown that when γ is a geodesic from P, then T_γ has a natural norm (Proposition 4.21), and that this normed vector space is a natural model for $T_\gamma^{\cdot}(T_P M)$ (see (6.11) ff.).

(4.14.2) Once Theorem 4.14.2 is proved, the notation of (4.11.5) -- (4.11.9) will apply to Jacobi fields of type $\underline{J}\{J\}$.

(4.15) Proof of Theorem 4.13. Let $0 < \nu^* < \nu$ and set

(4.15.1) $\tilde{\underline{J}} = \underline{J}_{\nu^*} - \underline{J}_\nu$ (see Figure 8);

then $\tilde{\underline{J}}$ is a Jacobi field along $\gamma\restriction[\nu,1]$ which vanishes at $X = \gamma(1)$. By Corollary 4.12.10, for $\mu \in [\nu,1]$,

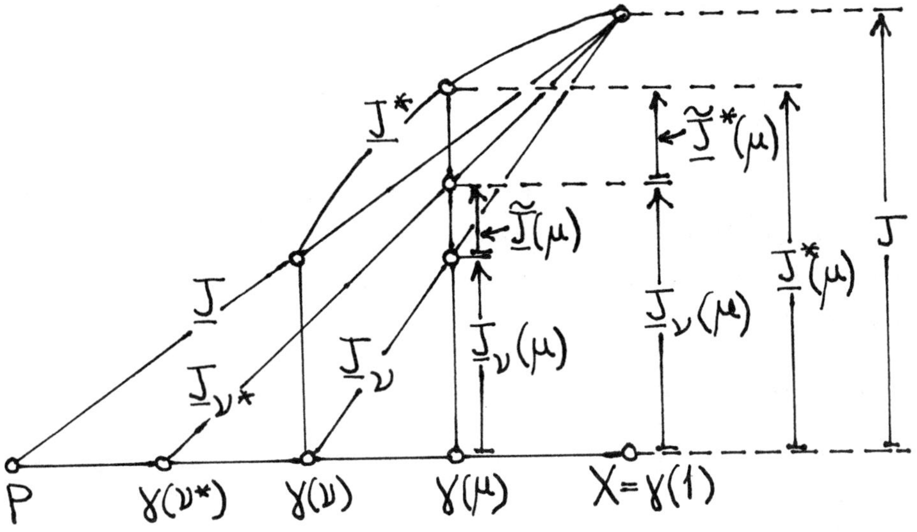

Figure 8

(4.15.2) $|\tilde{\underline{J}}(\mu)| \leq (3/2)((1-\mu)/(1-\nu))|\tilde{\underline{J}}(\nu)|$.

Now

(4.15.3) $|\tilde{\underline{J}}(\nu)| = |\underline{J}_{\nu*}(\nu)| \leq (3/2)\nu|J|$,

by Proposition 4.12.1. Substituting this in (4.15.2) gives

$$\begin{aligned}|\tilde{\underline{J}}(\mu)| &\leq (9/4)(1-\mu)(\nu/(1-\nu))|J| \\ &= (1-\mu)O(\nu;\ \text{fixed } |J|) \\ &= O(\nu;\ \text{fixed } |J|).\end{aligned}$$

From this and (4.15.1), (1) follows.

Proof of (2). In view of Corollary 4.12.9, given ν, there exists a unique Jacobi field $\underline{J}^*$ along $\gamma \restriction [\nu,1]$ such that

(4.15.4) $\underline{J}^*(\nu) = \underline{J}(\nu)$, $\underline{J}^*(1) = \underline{J}(1) = J$

(see Hicks [8, p. 147]). For $0 < \nu^* < \nu$, set

(4.15.5) $\tilde{\underline{J}}^* = \underline{J}^* - \underline{J}_{\nu*}$. (See Figure 8.)

The argument used to prove (4.15.2) applies to show that, for $\mu \in [\nu,1]$,

$$|\tilde{\underline{J}}^*(\mu)| \leq (3/2)((1-\mu)/(1-\nu))|\tilde{\underline{J}}(\nu)|.$$

Let $\nu^* \to 0^+$; then in the limit, by (4.15.5) and (1),

$$\begin{aligned}|\underline{J}^* - \underline{J}|(\mu) &\leq (3/2)((1-\mu)/(1-\nu))|\underline{J}^* - \underline{J}|(\nu) \\ &= 0,\end{aligned}$$

by (4.15.4). That is, $\underline{J} \restriction [\nu,1]$ is the Jacobi field $\underline{J}^*$. Hence $\underline{J}$ is a Jacobi field along $\gamma \restriction (0,1]$. It follows directly from the definition of $\underline{J}(1)$ that $\underline{J}$ has final condition $\underline{J}(1) = J$. In Proposition 4.12.1, set $\lambda = 1$ and let $\nu \to 0^+$; then in the limit,

(4.15.6) $|\underline{J}(\mu)| \leq (3/2)\mu|J|$,

which proves that $\lim_{\mu \to 0^+} |\underline{J}(\mu)| = 0$. Thus $\underline{J}$ satisfies (4.11.4). This completes the proof of (2).

Proof of (3). Let $\underline{J}^{\#}$ be a second Jacobi field along γ from P with final condition J. Let $\mu \in (0,1]$, and pick $\nu \in (0,\mu)$. Then Corollary 4.12.10 applies to $\underline{J}^{\#} - \underline{J}$ to yield

$$|\underline{J}^{\#} - \underline{J}|(\mu) \leq (3/2)((1-\mu)/(1-\nu))|\underline{J}^{\#} - \underline{J}|(\nu)$$
$$\leq (3/2)((1-\mu)/(1-\nu))(|\underline{J}^{\#}(\nu)| + |\underline{J}(\nu)|).$$

Let $\nu \to 0^+$; then $|\underline{J}^{\#}(\nu)|$ and $|\underline{J}(\nu)| \to 0$. It follows that $\underline{J}^{\#}(\mu) = \underline{J}(\mu)$; and since μ was arbitrary in $(0,1]$, this proves (3). q.e.d.

COROLLARY 4.15.7. In the notation of (4.14), for $\mu \in (0,1]$,

$$|\underline{J}\{J\}(\mu)| \leq (3/2)\mu|J|.$$

(1.6.3,VII) is not needed for this result.

Proof. This is a restatement of (4.15.6). q.e.d.

LEMMA 4.16. In the notation of Theorem 4.13, $(D^M_\gamma,\underline{J}_\nu)(1)$ converges to $(D^M_\gamma,\underline{J})(1)$ as $\nu \to 0^+$.

Proof. Let $\underline{J}$ be as in (4.15.1). Then

$$|D^M_\gamma,\tilde{\underline{J}}|(1) \leq (3/2)(1-\nu)^{-1}|\tilde{\underline{J}}(\nu)|, \quad \text{by Corollary 4.12.10}$$
$$\leq (9/4)(\nu/(1-\nu))|J|, \quad \text{by (4.15.3)}$$
$$= O(\nu;\ \text{fixed } |J|).$$

It follows that $\psi = \lim_{\nu\to 0^+} (D^M_\gamma,\underline{J}_\nu)(1)$ exists.

Now the $\underline{J}_\nu$, being Jacobi fields along γ, are solutions of the second order o.d.e. (4.11.2), and are therefore determined by their final conditions $J = \underline{J}_\nu(1)$ and $(D^M_\gamma,J_\nu)(1)$. Since solutions of (4.11.2) depend continuously on their final conditions, and since $\lim_{\nu\to 0^+} \underline{J}_\nu$ exists as a solution of (4.11.2), by Theorem 4.13.2, it follows that

$$\underline{J} = \lim_{\nu\to 0^+} \underline{J}_\nu \quad \text{has final conditions}$$

$$\lim_{\nu\to 0^+} (J,(D^M_\gamma,\underline{J}_\nu)(1)) = (J,\psi).$$

Therefore $\psi = (D^M_\gamma,\underline{J})(1)$. q.e.d.

LEMMA 4.17. Let $X \in B^{o}(P,M:\underline{\tau}^{\#})$, where $\underline{\tau}^{\#}$ is as in (4.4). Let γ_X be the geodesic from P to X, as in (3.4), and let $\tau(X)$ be as in (2.15.4). Let $\underline{J}\{J\}$ be a Jacobi field along γ_X from P, as in (4.14). Then

$$|D^M_{\gamma_X{'}1}\,\underline{J}\{J\} - J| = O((\tau X)^{2k})\,|J|.$$

This result requires (1.6.3,VII).

Proof. Let $\underline{J}_\nu$ be as in (4.11.3) with final condition J. Let $\underline{\Phi}_\nu$ be the parallel vector field along $\gamma_X \upharpoonright [\nu,1]$ such that

(4.17.1) $\underline{\Phi}_\nu(\nu) = (D^M_{\gamma_X{'}}\,\underline{J}_\nu)(\nu);$

and define the vector field $\underline{\Phi}_\nu$ along $\gamma_X \upharpoonright [\nu,1]$ by

$$\underline{\Psi}_\nu(\lambda) = (\underline{J}_\nu - (\lambda-\nu)\underline{\Phi}_\nu)(\lambda).$$

Note that

(4.17.2) $\underline{\Psi}_\nu(\nu) = 0.$

Then since $D^M_{\gamma_X{'}}\underline{\Phi}_\nu \equiv 0,$

(4.17.3) $(D^M_{\gamma_X{'}}\underline{\Psi}_\nu)(\mu) = (D^M_{\gamma_X{'}}\,\underline{J}_\nu - \underline{\Phi}_\nu)(\mu),$

for $\mu \in [\nu,\lambda]$. In particular,

(4.17.4) $(D^M_{\gamma_X{'}}\underline{\Psi}_\nu)(\nu) = 0.$

Also

$$((D^M_{\gamma_X{'}})^2\underline{\Psi}_\nu)(\mu) = ((D^M_{\gamma_X{'}})^2\underline{J}_\nu)(\mu)$$

$$= (R^M(\gamma_X{'},\underline{J}_\nu)\gamma_X{'})(\mu),$$

by (4.11.2). Now $|\gamma_X{'}| = \tau(X)$, and

$$d^M(P,\gamma_X(\mu)) = \mu|\gamma'| = \mu\tau(X).$$

Hence

$$\begin{aligned}|(D^M_{\gamma_X'}|D^M_{\gamma_X'}\underline{\Psi}_\nu|)|(\mu) &\le |(D^M_{\gamma_X'})^2\underline{\Psi}_\nu|(\mu)\\ &= |R^M(\gamma_X',\underline{J}_\nu)\gamma_X'|(\mu), \qquad \text{by (4.17.5)}\\ &= |\gamma_X'|^2|\underline{J}_\nu|(\mu)O((\mu\tau(X))^{-2+2k}), \text{ by Lemma 4.1.3,}\\ &= \tau(X)^{2k}|\underline{J}_\nu|(\mu)O(\mu^{-2+2k}; \text{ fixed } \tau(X)).\end{aligned}$$

Since

$$|\underline{J}_\nu(\mu)| \le (3/2)\mu|J|,$$

by Proposition 4.12.1, it follows that

$$|(D^M_{\gamma_X'}|D^M_{\gamma_X'}\underline{\Psi}_\nu|)|(\mu) = \tau(X)^{2k}|J|O(\mu^{-1+2k}; \text{ fixed } \tau(X)).$$

Integrating with respect to μ from ν to λ and using (4.17.4) gives

$$\begin{aligned}(4.17.6)\qquad |D^M_{\gamma_X'}\underline{\Psi}_\nu|(\lambda) &= \tau(X)^{2k}|J|O(\lambda^{2k}; \text{ fixed } \tau(X))\\ &= O(\tau(X)^{2k})|J|,\end{aligned}$$

since $\tau(X) < \underline{\tau}^\#$. Now since

$$|(D^M_{\gamma_X'}|\underline{\Psi}_\nu|)|(\lambda) \le |D^M_{\gamma_X'}\underline{\Psi}_\nu|(\lambda),$$

a further integration with respect to λ from ν to 1 yields, in view of (4.17.2),

$$|\underline{\Psi}_\nu|(1) = O(\tau(X)^{2k})|J|.$$

By definition of $\underline{\Psi}_\nu$, it follows that

$$\begin{aligned}|\underline{\Phi}_\nu(1) - (1-\nu)^{-1}J| &= (1-\nu)^{-1}O(\tau(X)^{2k})|J|\\ &= O(\tau(X)^{2k}; \text{ fixed } \underline{\nu})|J|,\end{aligned}$$

once ν is constrained to range over $(0,\underline{\nu}]$ for some fixed $\underline{\nu} < 1$. From

(4.17.6) and (4.17.3), with $\mu = \lambda = 1$,

$$|(D^M_{\gamma_X}, \underline{J}_\nu)(1) - \underline{\Phi}_\nu(1)| = O(\tau(X)^{2k};\ \text{fixed } \underline{\nu})|J|.$$

Therefore

$$(4.17.7)\quad |(D^M_{\gamma_X}, \underline{J}_\nu)(1) - (1-\nu)^{-1}J| = O(\tau(X)^{2k};\ \text{fixed } \underline{\nu})|J|.$$

As $\nu \to 0^+$, $(D^M_{\gamma_X}, \underline{J}_\nu)(1) \to (D^M_{\gamma_X}, \underline{J}\{J\})(1)$, by Lemma 4.16. In the limit, the left-hand side of (4.17.7) no longer depends on $\underline{\nu}$; so the lemma follows.

q.e.d.

LEMMA 4.18. Let M and P be as in (4.2) and (4.4). Let g be a geodesic from P with domain $[0,1]$ and let $J^\lambda \in T_{g\lambda}M$. Set $\tau^* = |g'| = d^M(P,g1)$. Then there is a unique Jacobi field $\underline{J}$ along g from P such that $\underline{J}(\lambda) = J^\lambda$. Moreover, for $\mu \in (0,1]$,

$$|\underline{J}(\mu)| = (\mu/\lambda)(1 + O(\tau^{*1/q})|J^\lambda|,$$

where $2/3 \leq 1 + O(\tau^{*1/q}) \leq 3/2$.

Proof. The uniqueness of $\underline{J}$ follows from Proposition 4.9 and Theorem 4.13.3 In view of (4.11.9) and (4.11.7) it is enough to construct $\underline{J}$ as required in case $\langle J^\lambda, g'\lambda\rangle = 0$. By Lemma 4.10 and (2.7.1), g has the form $g(\mu) = (Z^*, \mu\tau^*)$, for some $Z^* \in L$. There exists a C^∞ curve $Z^*(\sigma)$ through Z^* in L with domain $[-\underline{\sigma}, \underline{\sigma}]$ (for some $\underline{\sigma} > 0$) such that

$$Di_{\lambda\tau^*}\cdot Z^{*\prime}(\sigma)|_{\sigma=0} = \frac{d}{d\sigma}(Z^*\sigma, \lambda\tau^*)|_{\sigma=0} = J^\lambda.$$

(See (1.16.2) for i_t.) Define a vector field $\underline{J}$ along $g\restriction(0,1]$ by

$$(4.18.1)\quad \underline{J}(\mu) = Di_{\mu\tau^*}\cdot Z^{*\prime}(\sigma)|_{\sigma=0} = \frac{d}{d\sigma}(Z^*\sigma, \mu\tau^*)|_{\sigma=0};$$

and for $\sigma \in [-\underline{\sigma}, \underline{\sigma}]$, set

$$g_\sigma(\mu) = (Z^*\sigma, \mu\tau^*).$$

Then each g_σ is a geodesic from P by Lemma 4.10. Hence for each $\nu \in (0,1]$, $\{g_\sigma \upharpoonright [\nu,1]$ for $\sigma \in [-\underline{\sigma},\underline{\sigma}]\}$ is a one-parameter variation of g of class C^∞ through geodesics in M^o. Therefore $\underline{J}$ is a Jacobi field on $[\nu,1]$ for every $\nu \in (0,1]$. It follows that $\underline{J}$ is a Jacobi field along $g \upharpoonright (0,1]$.

Clearly $\underline{J}(\lambda) = J^\lambda$. By (4.18.1) and (4.4.2),

$$|\underline{J}(\mu)| = r(\mu\tau^*)|Z^{*\prime}0|.$$

When $\mu = \lambda$, this shows that

$$|\underline{J}^\lambda| = r(\lambda\tau^*)|Z^{*\prime}0|;$$

hence

$$(4.18.2) \qquad |\underline{J}(\mu)| = (r(\mu\tau^*)/r(\lambda\tau^*))|J^\lambda|.$$

By Lemma 4.3.1,

$$\begin{aligned} r(\mu\tau^*) &= \mu\tau^*(1 + O((\mu\tau^*)^{1/q})) \\ &= \mu\tau^*(1 + O(\tau^{*1/q})), \end{aligned}$$

and similarly for $r(\lambda\tau^*)$. Substituting into (4.18.2) gives

$$|\underline{J}(\mu)| = (\mu/\lambda)(1 + O(\tau^{*1/q}))|J^\lambda|.$$

By Lemma 4.3.3

$$\begin{aligned} (2/3)(\mu/\lambda) &= \frac{(4/5)\mu\tau^*}{(6/5)\lambda\tau^*} \\ &\le \frac{r(\mu\tau^*)}{r(\lambda\tau^*)} \\ &\le \frac{(6/5)\mu\tau^*}{(4/5)\lambda\tau^*} \\ &= (3/2)(\mu/\lambda). \end{aligned}$$

When substituted into (4.18.2), this completes the proof of the lemma.

q.e.d.

PROPOSITION 4.19. Let γ be a <u>geodesic</u> from P, satisfying (4.11.12). Let $\nu < \lambda \le 1$, let $\underline{J}_\nu$ be a Jacobi field along $\gamma \upharpoonright [\nu,1]$ satisfying (4.11.3), and set $J_\nu^\lambda = \underline{J}_\nu(\lambda)$. Then:

(1) For $\mu \in [\nu,\lambda]$,

$$|\underline{J}_\nu(\mu)| \geq ((\mu-\nu)/(\lambda-\nu))(1 + O((\lambda T)^{1/q}))|J^\lambda_\nu|$$
$$\geq (2/3)((\mu-\nu)/(\lambda-\nu))|J^\lambda_\nu|;$$

(2) For $\mu \in [\lambda,1]$,

$$|\underline{J}_\nu(\mu)| \leq ((\mu-\nu)/(\lambda-\nu))(1 + O(T^{1/q}))|J^\lambda_\nu|$$
$$\leq (3/2)((\mu-\nu)/(\lambda-\nu))|J^\lambda_\nu|;$$

(3) $|D^M_{\gamma'}\underline{J}_\nu|(\nu) \geq (2/3)(1/(\lambda-\nu))|J^\lambda_\nu|$.

Proof. By (4.11.9) and (4.11.7) it is enough to prove the lemma when $\underline{J}_\nu$ is orthogonal to γ'. Let M be constructed according to (4.4) subject to (4.2.4) -- (4.2.6). Let g be a geodesic in M from P with domain $[0,1]$ such that $|g'| = |\gamma'|$; since this is $<\underline{\tau}^\#$ by (4.11.12), such a g does exist. Let $\underline{J}$ be a Jacobi field along g from P such that $\underline{J}$ is orthogonal to g' and $|\underline{J}(\lambda-\nu)| = |J^\lambda_\nu|$. Let $\Pi^M(\mu)$ and $\Pi^M(\mu-\nu)$ be the 2-planes spanned respectively by $\gamma'(\mu)$ and $\underline{J}_\nu(\mu)$ in $T_{\gamma\mu}M$, and by $g'(\mu-\nu)$ and $J(\mu-\nu)$ in $T_{g(\mu-\nu)}M$. Then

$$\text{(4.19.4)}\quad \begin{aligned} d^M(P,\gamma\mu) &= \mu|\gamma'|, \quad \text{by (2.7.1)}\\ &> (\mu-\nu)|g'|\\ &= d^M(P,g(\mu-\nu)). \end{aligned}$$

Hence

$$\begin{aligned} K^M(\Pi^M(\mu)) &> -A(\mu|\gamma'|)^{-2+1/q}, &&\text{by (4.2.6)}\\ &> -A((\mu-\nu)|g'|)^{-2+1/q}, &&\text{by (4.19.4)}\\ &= K^M(\Pi^M(\mu-\nu)), \end{aligned}$$

by Lemma 4.8.2. As in the proof of Proposition 4.12, Rauch's comparison theorem may be applied to $\gamma\restriction[\nu,1]$ and $g\restriction[0,1-\nu]$. As analogue of (4.12.4) we obtain

$$|\underline{J}_\nu(\mu)| \begin{cases} \geq |\underline{J}_\nu(\mu-\nu)|, & \text{for } \mu \in [\nu,\lambda],\\ \leq |\underline{J}_\nu(\mu-\nu)|, & \text{for } \mu \in [\lambda,1]. \end{cases}$$

The proof of the proposition is now analogous to the rest of the proof of Proposition 4.12; the inequality signs are reversed, and (4.12.5) -- (4.12.7) are replaced by Lemma 4.18. q.e.d.

COROLLARY 4.19.5. Let γ be a geodesic from P satisfying (4.11.12) and let $\underline{J}$ be a Jacobi field along γ from P. Then, for $\lambda,\mu \in (0,1]$,

$$|\underline{J}(\mu)| = (\mu/\lambda)(1 + O(T^{1/q}))|\underline{J}(\lambda)|,$$
$$\text{where } 1/2 \leq 1 + O(T^{1/q}) \leq 3/2.$$

When $\mu \leq \lambda$, $1 + O(T^{1/q})$ can be improved to $1 + O((\lambda T)^{1/q})$.

Proof. For every $\nu \in (0,1)$, let $\underline{J}_\nu$ be the Jacobi field along $\gamma\restriction[\nu,1]$ from $\gamma(\nu)$ with the same final condition that $\underline{J}$ has. By Propositions 4.12 and 4.19, once $\nu < \lambda$ and μ,

$$|\underline{J}_\nu(\mu)| = ((\mu-\nu)/(\lambda-\nu))(1 + O(T^{1/q}))|\underline{J}_\nu(\lambda)|;$$

and the riders stated above hold for $1 + O(T^{1/q})$. Taking the limit as $\nu \to 0^+$ and applying Theorem 4.13 gives the corollary . q.e.d.

LEMMA 4.20. Let $X \in B^o(P,M;\underline{T}^{\#})$, and let γ_X be the geodesic from P to X, as in (3.4). For each $\psi \in T_XM$, let $\underline{J}\{\psi\}$ be the Jacobi field of (4.14) along γ_X from P with final condition ψ. Then for each $\lambda \in (0,1]$ the rule $\psi \to \underline{J}\{\psi\}(\lambda)$ defines a linear isomorphism

$$\lambda_*\colon\ T_XM \to T_{\gamma_X\mu}M.$$

Moreover for all ψ,

$$|\lambda_*(\psi)| = \lambda(1 + O(\underline{T}^{\#1/q}))|\psi|,$$

where $1/2 \leq 1 + O(\underline{T}^{\#1/q}) \leq 3/2$.

Proof. For each $\nu \in (0,\lambda)$ let $\underline{J}\{\psi\}_\nu$ be the Jacobi field along $\gamma_X\restriction[\nu,1]$ satisfying (4.11.3) and with final condition ψ. The rule

(4.20.1) $\psi \to \underline{J}\{\psi\}_\nu(\lambda)$

is a linear transformation. Since $\gamma_X(\nu)$ has no conjugate points along γ_X between $\gamma_X(\lambda)$ and X, by Corollary 4.12.9, the rule (4.20.1) is an isomorphism $T_X M \approx T_{\gamma_X} \lambda M$. Let $\nu \to 0^+$; then λ_* is the limit of these isomorphisms, and is therefore a linear transformation. The assertion about $|\lambda_*(\psi)|$ follows from Corollary 4.19.5, with $T = \underline{\tau}^\#$. It follows that $\ker(\lambda_*) = 0$; and since $\dim(T_X M) = \dim(T_{\gamma_X \mu} M)$, λ_* is an isomorphism. q.e.d.

PROPOSITION 4.21. Let $X \in B^o(P,M;\underline{\tau}^\#)$. For each $\psi \in T_X M$, let $\underline{J}\{\psi\}$ be as in (4.14). Then:

(1) $|\underline{J}\{\psi\}|_P = \lim_{\lambda\to 0^+} |\underline{J}\{\psi\}(\lambda)|/\lambda$ exists;

(2) $|\underline{J}\{\psi\}|_P = (1 + O(\underline{\tau}^{\#1/q}))|\psi|$, where $1/2 \le 1 + O(\underline{\tau}^{\#1/q}) \le 3/2$;

(3) $|\ |_P$ is a norm on T_{γ_X} (of (4.14)). (See Figure 11, p. 132.)

Proof of (1). I abbreviate $\underline{J}\{\psi\}$ to $\underline{J}$ in this part of the proof. By Corollary 4.19.5, with λ replaced by 1 and μ by λ.

(4.21.4 $|\underline{J}(\lambda)|/\lambda \le (3/2)|\psi|$.

By the same corollary, with $T = \underline{\tau}^\#$ and with $\mu \in (0,\lambda)$,

$$|\underline{J}(\mu)|/\mu - |\underline{J}(\lambda)|/\lambda = O((\lambda\underline{\tau}^\#)^{1/q})|\underline{J}(\lambda)|/\lambda$$
$$= O(\lambda^{1/q};\ \text{fixed } \underline{\tau}^\#)|\psi|,$$

by (4.21.4). The existence of $|\underline{J}\{\psi\}|_P$ follows.

Proof of (2). This follows from (1) and Lemma 4.20 by taking the limit as $\lambda \to 0^+$.

Proof of (3). For each $\lambda \in (0,1]$, the rule

$$\psi \to |\underline{J}\{\psi\}(\lambda)|/\lambda$$

is a norm on $T_X M$, by Lemma 4.20. Taking the limit as $\lambda \to 0^+$ shows that

$$|\underline{J}\{\psi_1 + \psi_2|_P \le |\underline{J}\{\psi_1\}|_P + |\underline{J}\{\psi\}_2|_P, \quad \text{and}$$
$$|\underline{J}\{r\psi\}|_P = |r|\,|\underline{J}\{\psi\}|_P, \quad \text{for all } r \in \mathbb{R}.$$

From (2),

$$|\underline{J}\{\psi\}|_P = 0 \text{ if and only if } \psi = 0.$$

Thus $| \, |_P$ is a norm. q.e.d.

(4.22) Example. If ψ is a scalar multiple of $\gamma_X'(1)$, then $|\underline{J}\{\psi\}|_P = |\psi|$. For $\underline{J}\{\psi\} = {}^{\|}\underline{J}\{\psi\}$ (see (4.11.7)); and $|{}^{\|}\underline{J}\{\psi\}(\lambda)|$ is linear in λ by (4.11.7).

(4.23) Example. If M is a smooth Riemannian manifold at P, then to $\psi \in T_X M$ corresponds a unique $\tilde{\psi} \in T_{\tilde{X}}(T_P M)$, where $\tilde{X} = \exp_P^{-1}(X)$, such that $\mathrm{Dexp}_P(X) \cdot \tilde{\psi} = \psi$. Now $|\underline{J}\{\psi\}|_P = |\tilde{\psi}|$.

(4.24) Example. Let M be as in (4.2) and (4.4), let $g = g_{(Z^*,\tau^*)}$, and let $v \in T_{(Z^*,\tau^*)}M$. Then $| \, |_P$ is determined on T_g by: (4.23), in case v is a scalar multiple of $g'(1)$; and by

(4.24.1) $|\underline{J}\{v\}|_P = (\tau^*/r(\tau^*))|v|$, in case $\langle g'(1), v\rangle = 0$.

For it follows from (4.18.2), in the latter case, that

$$|\underline{J}\{v\}(\mu)|/\mu = (r(\mu\tau^*)/\mu\tau^*)(\tau^*/r(\tau^*))|v|;$$

and since

$$\lim_{\mu\to 0^+} r(\mu\tau^*)/(\mu\tau^*) = \lim_{\mu\to 0^+} r'(\mu\tau^*) = 1,$$

(4.24.1) follows.

CHAPTER 5: ONE-PARAMETER FAMILIES OF GEODESICS FROM P

Introduction

The purpose of this chapter is to prove that the results of Chapter 4 about Jacobi fields from P can be "integrated" to give results about 1-parameter families of geodesics from P. Let β be a C^∞ path in $B^o(P,M;\underline{\tau})$. Assuming $\underline{\tau}$ is small enough (see (5.12)), then for each σ, there is a unique geodesic $\gamma_{\beta\sigma}$ from P to $\beta(\sigma)$ with domain [0,1], by Theorem 3.3. Theorem 5.13 states that the family $\gamma_{\beta\sigma}$ is differentiable with respect to σ and that for each σ, $\underline{\partial}_\sigma \gamma_{\beta\sigma}$ is the Jacobi field $\underline{J}[0]$ along $\gamma_{\beta\sigma}$ from P with final condition $\beta'(\sigma)$ and "initial condition 0".

In proving this theorem I may assume $\sigma = 0$, and I may then restrict attention to a (short) initial segment of β. The first step (Proposition 5.2 and (5.3)) is to show that whenever $\nu > 0$ is small enough there is a 1-parameter family of weak geodesics $\gamma_{\nu,\sigma}$ from $\gamma_{\beta 0}(\nu)$ to $\beta(\sigma)$. This is done by applying a covering-homotopy argument to lift β through the immersion $\exp_{\gamma_{\beta 0}\nu}$: $T_{\gamma_{\beta 0}\nu}M \to M^o$ (defined near $\gamma_{\beta 0}(\nu)$). Set $\underline{\Delta}(\nu,\lambda,\sigma) = \gamma_{\nu,\sigma}(\lambda)$. The second step is to examine the convergence of $\underline{\Delta}$ as $\nu \to 0^+$. Corollary 5.8.9 shows that $\Delta_0(\lambda,\sigma) = \lim_{\nu\to 0^+} \underline{\Delta}(\nu,\lambda,\sigma)$ exists. Since, for fixed σ, each $\gamma_{\nu,\sigma}$ is a weak geodesic, $\lim_{\nu\to 0^+} \gamma_{\nu,\sigma}$ is a weak geodesic from P to $\beta(\sigma)$. Proposition 3.12 shows that this limiting path must in fact be $\gamma_{\beta\sigma}$; that is, $\Delta_0(\lambda,\sigma) = \gamma_{\beta\sigma}(\lambda)$ (Proposition 5.9). Now for $\sigma = 0$, $\underline{\Delta}(\nu,\lambda,0) = \gamma_{\beta 0}(\lambda)$ is constant in ν. This makes it possible to show that $\lim_{\nu\to 0^+} \underline{\partial}_\sigma \underline{\Delta}(\nu,\lambda,0)$ exists, and that $\underline{\partial}_\sigma \Delta_0(\lambda,0)$ also exists and equals this limit (Lemma 5.10). But $\underline{\partial}_\sigma \underline{\Delta}(\nu,\lambda,0) = \underline{J}_\nu[0](\lambda)$, where $\underline{J}_\nu[0]$ is the Jacobi field along $\gamma_{\beta 0}\restriction[\nu,1]$ with terminal values $\underline{J}_\nu[0](\nu) = 0$, $\underline{J}_\nu[0](1) = \beta'(0)$. Since $\lim_{\nu\to 0^+} \underline{J}_\nu[0] = \underline{J}[0]$ by Theorem 4.13, this proves Theorem 5.13.

Theorem 5.13 implies that given a (short enough) geodesic γ_X from P, a Jacobi field $\underline{J}[0]$ along γ_X from P, and a C^∞ curve β from X such that

$\beta'(0) = \underline{J}[0](1)$, then the 1-parameter family $\gamma_{\beta\sigma}$ of geodesics from P is an "integral" of $\underline{J}[0]$ with "final condition" β.

Two refinements of Theorem 5.13 will be needed in Chapter 6. One is that β need not be C^∞ but only differentiable (Proposition 5.15). The other is that instead of being given the final condition β, one can be given an "intermediate condition": for some $\underline{\lambda} \in (0,1)$, a path $\beta^{\underline{\lambda}}$ from $\gamma_X(\underline{\lambda})$ such that $\beta^{\underline{\lambda}}{}'(0) = \underline{J}[0](\underline{\lambda})$ (Proposition 5.17).

Hypothesis VII of (1.6.3) is used only in the last result of this chapter, Lemma 5.18. This is not used in proving the main theorem, Theorem 6.9; it will show that $\exp_P\colon T_PM \to M$ near P satisfies (1.6.3,V) (see Theorem 6.13).

Covering Homotopies

(5.1) Notation. Let h be a fixed chart for M at P. Given $X \in M^o$, let $\tau(X)$ be as in (2.15.4) and γ_X as in (3.4). Let $\underline{r}$ be as in (3.13.1), and let $\underline{\tau}^{\#}$ satisfy (3.13.2), Lemma 4.3 and (4.4.1); these quantities depend on the choice of h. From (3.9.2),

(5.1.1) $B(P,M;\underline{\tau}^{\#}) \subseteq \mathrm{im}(h)$.

Set

(5.1.2) $\hat{\underline{\tau}} = \underline{\tau}^{\#}/(1 + (3/2)\underline{r})$,

and assume henceforth that

(5.1.3) $\tau(X) < \hat{\underline{\tau}}$.

By (3.13.1) and (3.9.1), $\underline{r} < 1/16$; hence

(5.1.4) $B(X,M;\underline{r}\tau(X)) \subseteq \mathrm{im}(h) \cap B^o(P,M;\underline{\tau}^{\#})$.

Let $\beta\colon [0,1] \to M^o$ be a C^∞ path such that

(5.1.5) $\beta(0) = X$, and $L(\beta) < \underline{r}\tau(X)$;

and set

(5.1.6) $B = \sup\{|\beta'(\sigma)| \text{ for } \sigma \in [0,1]\}$.

(5.1.7) ν will denote a number in $(0,1/4)$.

Given such a ν, set

(5.1.8) $N = N(\nu) = \gamma_X(\nu)$.

Then

(5.1.9) $\tau(N) = \nu\tau(X)$.

As in (3.6), let $\tilde{U}_N \subseteq T_N M$ be the natural domain of the exponential map e_N of M at N.

PROPOSITION 5.2. With the notation of (5.1), there is a unique continuous path $\tilde{\beta}_\nu$: $[0,1] \to \tilde{U}_N$ such that:

(1) $\tilde{\beta}_\nu(0) = (1-\nu)\gamma_X'(\nu)$;

(2) $e_N \circ \tilde{\beta}_\nu = \beta$;

(3) $|\tilde{\beta}_\nu(\sigma)| \leq (1-\nu)\tau(X) + (3/2)L(\beta \restriction [0,\sigma])$.

Moreover, (4) $\tilde{\beta}_\nu$ is C^∞.

(5.3) Notation. Whenever $\tilde{\beta}_\nu(\sigma)$ is defined in Proposition 5.2, set

(5.3.1) $\tilde{\Delta}_\nu(\lambda,\sigma) = \tilde{\beta}_\nu^\lambda(\sigma) = \tilde{\gamma}_{\nu,\sigma}(\lambda) = ((\lambda-\nu)/(1-\nu))\tilde{\beta}_\nu(\sigma)$.

Also set

(5.3.2) $\Delta_\nu(\lambda,\sigma) = \beta_\nu^\lambda(\sigma) = \gamma_{\nu,\sigma}(\lambda) = e_N \circ \tilde{\Delta}_\nu(\lambda,\sigma)$. (See Figure 9.)

Note that $\tilde{\beta}_\nu^1 = \tilde{\beta}$, $\beta_\nu^1 = \beta$, and that

(5.3.3) $\tilde{\gamma}_{\nu,\sigma}(\nu) = N = \gamma_X(\nu) = \gamma_{\nu,\sigma}(\nu)$.

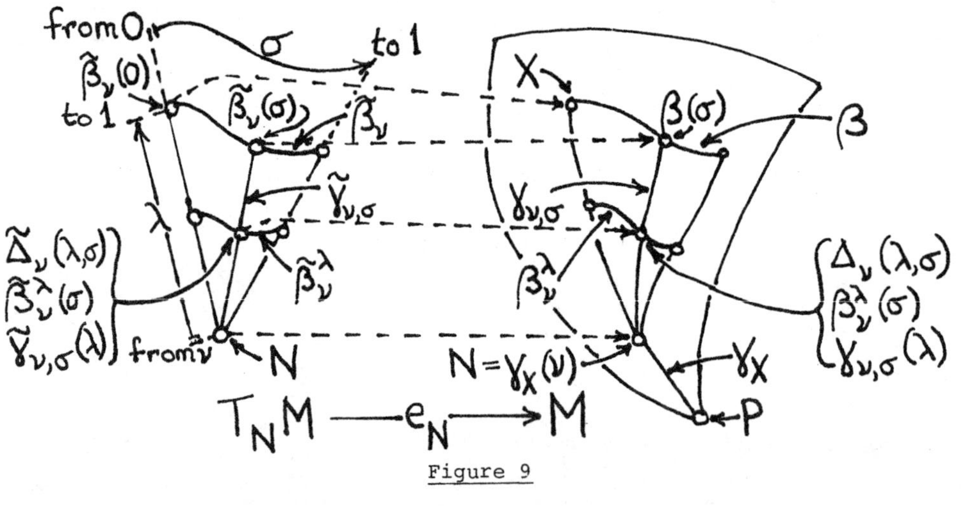

Figure 9

Once Proposition 5.2 is proved, this notation will define C^∞ maps

$$\tilde{\Delta}_\nu\colon [\nu,1]] \times [0,1] \to T_N M, \quad \Delta_\nu\colon [\nu,1] \times [0,1] \to M^o, \quad \text{etc.}$$

(5.4) Proof of Proposition 5.2. The proof incorporates Lemma 5.5. To simplify the notation I shall omit the subscripts ν and N except in formulae to be quoted subsequent to Lemma 5.5.

Let Σ be the set of $\sigma^* \in [0,1]$ such that there exists a unique $\tilde{\beta}\colon [0,\sigma^*] \to \tilde{U}$ satisfying (1) -- (4). It suffices to show that Σ is non-empty, open and closed in [0,1]. Σ is non-empty, since $0 \in \Sigma$; (3) holds because $|\gamma_X'| = \tau(X)$, and (4) is vaccuous. Σ is open. Let $\sigma^* \in \Sigma$ with $\sigma^* < 1$. Then $[0,\sigma^*] \subseteq \Sigma$. Now

(5.4.1) $|\tilde{\beta}_\nu(\sigma)| < \underline{\tau}^\#$ whenever (3) holds for σ, by (5.1.2), (5.1.3) and (5.1.5)

$< \underline{\tau}^*$

of Proposition 3.7, by (3.13.2). By Proposition 3.7.2, e is regular at $\tilde{\beta}(\sigma)$ and

(5.4.2) $|\tilde{\beta}_\nu'(\sigma)| \leq (3/2)|\beta'(\sigma)|$, whenever (3) holds for σ

$\leq (3/2)B$,

by (5.1.6). In particular, returning to the context $[0,\sigma^*] \subseteq \Sigma$, since e is locally a diffeomorphism at $\tilde{\beta}(\sigma^*)$, there is a unique extension of $\tilde{\beta}$ to $[0,\sigma^{**})$ for some $\sigma^{**} \in (\sigma^*,1]$, such that on $[0,\sigma^{**})$, (2) holds and $|\tilde{\beta}(\sigma)| < \underline{\tau}^*$. Then (4) and (5.4.2) hold on $[0,\sigma^{**})$. Integrating the latter with respect to σ yields

$$L(\tilde{\beta}\restriction[0,\sigma]) \leq (3/2)L(\beta\restriction[0,\sigma]).$$

Also

$$\begin{aligned} |\tilde{\beta}(0)| &= (1-\nu)|\gamma_X'|, \qquad \text{by (1)} \\ &= (1-\nu)\tau(X). \end{aligned}$$

Hence

$$\begin{aligned} |\tilde{\beta}(\sigma)| &\leq |\tilde{\beta}(0)| + L(\tilde{\beta}\restriction[0,\sigma]) \\ &\leq (1-\nu)\tau(X) + ((3/2)L(\beta\restriction[0,\sigma]); \end{aligned}$$

that is, (3) holds on $[0,\sigma^{**})$. This proves that $[0,\sigma^{**}) \subseteq \Sigma$, which is therefore open.

The preceding argument also shows that

(5.4.3) whenever $\sigma^* \in [0,1]$ with $[0,\sigma^*] \subseteq \Sigma$, then $\tilde{\beta}$ is C^∞ on $[0,\sigma^*]$.

Σ is closed. Assume that $[0,\sigma^*) \subseteq \Sigma$. Then (5.4.2) holds on $[0,\sigma^*)$; hence $\lim_{\sigma\to\sigma^*_-} \tilde{\beta}(\sigma)$ exists in $T_N M$, and $\tilde{\beta}(\sigma^*)$ must perforce be this limit. Thus $\tilde{\beta}$ is now defined and continuous on $[0,\sigma^*]$. Assuming Lemma 5.5 (below), it follows that $\sigma^* \in \Sigma$. For (1) is clear, (2) and (3) at σ^* follow from their continuity at $\sigma \in [0,\sigma^*)$ as $\sigma \to \sigma^*_-$, and (4) holds by (5.4.3). Hence Σ is closed. This reduces Proposition 5.2 to Lemma 5.5.

LEMMA 5.5. In the preceding context, $\tilde{\beta}_\nu(\sigma^*) \in \tilde{U}_N$.

Proof. (I continue to omit the subscripts ν and N.) Let $\Lambda^* = \{\lambda \in [\nu,1]$ such that $\tilde{\Delta}(\sigma^*,\lambda) \in \tilde{U}\}$. It is sufficient to show that $\Lambda^* = [\nu,1]$. Now Λ^* is a non-empty initial segment of $[\nu,1]$ and is open in $[\nu,1]$, so it remains to show that Λ^* is closed.

Assume that $[\nu,\lambda^*) \subseteq \Lambda^*$. Then the domain of $\tilde{\Delta}$ contains

$$\mathcal{D} = [\nu,1] \times [0,\sigma^*) \cup [\nu,\lambda^*) \times \{\sigma^*\},$$

and

(5.5.1) $\tilde{\Delta}: \mathcal{D} \to \tilde{U}$.

Hence (see (3.6))

$$\Delta: \mathcal{D} \to M^o.$$

The first step is to show that Δ is uniformly continuous on $\mathcal{D}$. This follows from (5.5.2) and (5.5.4), which show that $\left|\frac{\partial}{\partial\lambda}\Delta(\lambda,\sigma)\right|$ and $\left|\frac{\partial}{\partial\sigma}\Delta(\lambda,\sigma)\right|$ are bounded above, uniformly in λ and σ.

(5.5.2)
$$\begin{aligned} |\partial_\lambda \Delta_\nu(\lambda,\sigma)| &= |\gamma_{\nu,\sigma}{}'(\lambda)| \\ &= |\tilde{\gamma}_{\nu,\sigma}{}'(\lambda)|, \text{ by the exponential property of e} \\ &= |\tilde{\beta}_\nu(\sigma)|/(1-\nu), \text{ by (5.3.1)} \\ &\le 2B, \qquad \text{by (5.4.2)}. \end{aligned}$$

For fixed $\sigma \in [0,\sigma^*)$, $\frac{\partial}{\partial\sigma}\Delta(\lambda,\sigma) = \frac{\partial}{\partial\sigma}\gamma_\sigma(\lambda)$ is a Jacobi field along γ_σ, which is a weak geodesic in M^o by (5.3.2). Now γ_σ has domain $[\nu,1]$, and therefore length

$$(5.5.3)\quad L(\gamma_{\nu,\sigma}) = (1-\nu)|\gamma_{\nu,\sigma}'| = |\tilde{\beta}_\nu(\sigma)|, \quad \text{by } (5.5.2)$$
$$< \underline{\tau}^{\#} \quad \text{by } (5.4.1)$$
$$< \pi/(2K^{1/2}), \quad \text{by } (3.9.2) \text{ and } (3.13.2).$$

So Proposition 4.12.1 gives

$$(5.5.4)\quad |\partial_\sigma \Delta_\nu(\lambda,\sigma)| \le (3/2)\lambda\left|\frac{\partial}{\partial\sigma}\Delta(1,\sigma)\right| = (3/2)\lambda|\beta'(\sigma)|, \quad \text{by } (5.3.2)$$
$$\le (3/2)\lambda B, \quad \text{by } (5.1.6)$$
$$\le (3/2)B.$$

It follows that Δ is uniformly continuous on $\mathcal{D}$.

Therefore

$$\Delta(\lambda^*,\sigma^*) = \lim_{\substack{(\lambda,\sigma)\to(\lambda^*,\sigma^*)\\ (\lambda,\sigma)\in\mathcal{D}}} \Delta(\lambda,\sigma)$$

exists. In particular,

$$(5.5.5)\quad \Delta(\lambda^*,\sigma^*) = \lim_{\sigma\to\sigma^*_-} \beta^{\lambda^*}(\sigma) = \beta^{\lambda^*}(\sigma^*),$$

say; and

$$(5.5.6)\quad \Delta(\lambda^*,\sigma^*) = \lim_{\lambda\to\lambda^*_-} \gamma_{\sigma^*}(\lambda) = \gamma_{\sigma^*}(\lambda^*),$$

say. Now, using (5.5.4)

$$(5.5.7)\quad |\beta_\nu^{\lambda^*\prime}(\sigma)| = \left|\frac{\partial}{\partial\sigma}\Delta_\nu(\lambda,\sigma)\right| \le (3/2)\lambda^*|\beta'(\sigma)|;$$

hence

$$(5.5.8)\quad L(\beta^{\lambda^*}\restriction[0,\sigma^*]) \le (3/2)\lambda^* L(\beta\restriction[0,\sigma^*]) < (3/2)\lambda^*\underline{r}\tau(X), \quad \text{by } (5.1.5)$$

$$< (1/4)\lambda^*\tau(X),$$

since $\underline{r} < 1/16$ (see (5.1)). Therefore

$$\begin{aligned} d^M(P,\beta^{\lambda^*}(\sigma^*)) &\geq d^M(P,\beta^{\lambda^*}(0)) - L(\beta^{\lambda^*}\restriction[0,\sigma^*]) \\ &> d^M(P,\gamma_X\lambda^*) - (1/4)\lambda^*\tau(X) \\ &= \lambda^*\tau(X) - (1/4)\lambda^*\tau(X) \\ &= (3/4)\lambda^*\tau(X) \\ &> 0. \end{aligned}$$

That is, $\beta^{\lambda^*}(\sigma^*) \in M^O$. So by (5.5.5) and (5.5.6),

(5.5.9) $\gamma_{\sigma^*}(\lambda^*) \in M^O$.

Now define γ: $[0,1] \to M$ by

$$\gamma(\mu) = \gamma_{\sigma^*}(\mu^{\lambda^*}).$$

Then $\mathrm{im}(\gamma) \subseteq M^O$, because $\gamma(1) \in M^O$ by (5.5.9), and for $\mu \in [0,1)$,

(5.5.10) $\gamma(\mu) = e\circ\tilde{\gamma}_{\sigma^*}(\mu\lambda^*) \in M^O$,

since $\tilde{\gamma}_{\sigma^*}(\mu\lambda^*) \in \tilde{U}$ by (5.5.1). Moreover (5.5.10) shows that γ is a weak geodesic from N in M^O. According to (3.6), this means that $\gamma'(0) = \gamma_{\sigma^*}(\lambda^*)$ is in $\tilde{U}$. That is, $\lambda^* \in \Lambda^*$; so Λ^* is closed. This proves Lemma 5.5, and with it, Proposition 5.2. q.e.d.

PROPOSITION 5.6. Define $\underline{\tilde{\beta}}$: $(0,1/4) \times [0,1] \to TM$ by $\underline{\tilde{\beta}}(\nu,\sigma) = \tilde{\beta}_\nu(\sigma)$, where $\tilde{\beta}_\nu$ is as in Proposition 5.2. Then $\underline{\tilde{\beta}}$ is of class C^∞.

Proof. I first show that $\underline{\tilde{\beta}}$ is continuous. By (5.4.2) the family $\{\tilde{\beta}_\nu$ for $\nu \in (0,1/4)\}$ is equicontinuous in σ, and by (5.4.1) it is uniformly bounded. By the Ascoli-Arzela theorem and the uniqueness of each $\tilde{\beta}_\nu$, guaranteed by Proposition 5.2, it follows that

$$\tilde{\beta}_{\nu^*} = \lim_{\nu\to\nu^*} \tilde{\beta}_\nu,$$

whenever $\nu \to \nu^*$ in $(0,1/4)$; and in fact, the convergence is uniform in σ.

From this fact and the equicontinuity of $\{\tilde{\beta}_\nu\}$ in σ, it follows that $\underline{\tilde{\beta}}$ is continuous in (ν,σ).

Set

$$\underline{N} = \{N(\nu),\ \text{for } \nu \in (0,1/4)\} = \mathrm{im}(\gamma_X \restriction (0,1/4)),$$

and set

$$T_{\underline{N}}M = \cup\{T_{N(\nu)}M,\ \text{for } \nu \in (0,1/4)\}.$$

Since $\gamma_X \restriction (0,1/4)$ is C^∞ by Proposition 2.13, $T_{\underline{N}}M$ is a C^∞ submanifold of $T(M^o)$. For each $N \in \underline{N}$, set

$$\tilde{V}_N = \tilde{U}_N \cap B(N,T_N M;\underline{\tau}^*),$$

where $\underline{\tau}^*$ is as in Proposition 3.7; and set

$$\underline{\tilde{V}} = \cup\{V_N,\ \text{for } N \in \underline{N}\} \subseteq T_{\underline{N}}M.$$

Then $\underline{\tilde{V}}$, being an open submanifold of $T_{\underline{N}}M$, is also a C^∞ submanifold of $T(M^o)$. Define $\exp_{\underline{N}}\colon\ \tilde{V}_N \to M^o$ by

$$\exp_{\underline{N}} \restriction \tilde{V}_N = e_N \restriction \tilde{V}_N.$$

Then $\exp_{\underline{N}}$ is of class C^∞, and is regular on each $\tilde{V}_N$ by Proposition 3,7,2, According to the implicit function theorem (see Dieudonné [4, p. 265]), given $(\nu^*,\sigma^*) \in (0,1/4) \times [0,1]$ there is a unique continuous function B defined in a neighbourhood of (ν^*,σ^*) with range in $\underline{\tilde{V}}$, such that $\exp_{\underline{N}} \circ B(\nu,\sigma) = \beta(\sigma)$ and $B(\nu^*,\sigma^*) = \underline{\tilde{\beta}}(\nu^*,\sigma^*)$; moreover B is of class C^∞. Since $\underline{\tilde{\beta}}$ itself has the properties postulated of B, B must be a restriction of $\underline{\tilde{\beta}}$. Hence $\underline{\tilde{\beta}}$ is of class C^∞. q.e.d.

One-Parameter Families of Geodesics

(5.7) Notation. Set

$$\underline{\mathcal{D}} = \{(\nu,\lambda,\sigma) \in (0,1/4) \times (0,1] \times [0,1] \text{ such that } \nu < \lambda\},$$

and define $\underline{\Delta}\colon\ \underline{\mathcal{D}} \to M^o$ by

(5.7.1) $\underline{\Delta}(\nu,\lambda,\sigma) = \Delta_\nu(\lambda,\sigma)$

(see (5.3.2)).

LEMMA 5.8. $\underline{\Delta}$ is C^∞, and:

(1) $|\partial_\lambda \underline{\Delta}(\nu,\lambda,\sigma)| = |\gamma_{\nu,\sigma}'(\lambda)| < \underline{\tau}^\#/(1-\nu)$;

(2) $|\partial_\sigma \underline{\Delta}(\nu,\lambda,\sigma)| = |\beta_\nu^\lambda{}'(\sigma)| \leq (3/2)\lambda|\beta'(\sigma)|$;

(3) $|\partial_\nu \underline{\Delta}(\nu,\lambda,\sigma)| \leq 4(1-\lambda)B\sigma$.

<u>Proof</u>. By (5.3.2),

$$\underline{\Delta}(\nu,\lambda,\sigma) = e_{N(\nu)}((\lambda-\nu)/(1-\nu))\underline{\tilde{\beta}}(\nu,\sigma).$$

Since $N(\nu) = \gamma_X(\nu)$ is C^∞ in ν by Proposition 2.13, and $\underline{\tilde{\beta}}$ is C^∞ in (ν,σ) by Proposition 5.6, it follows that $\underline{\Delta}$ is C^∞.

<u>Proof of (1)</u>.

$$\begin{aligned} |\partial_\lambda \underline{\Delta}(\nu,\lambda,\sigma)| &= |\gamma_{\nu,\sigma}'(\lambda)|, && \text{by (5.3.2)} \\ &= |\tilde{\beta}_\nu(\sigma)|/(1-\nu), && \text{by (5.5.2)} \\ &< \underline{\tau}^\#/(1-\nu), && \text{by (5.5.3).} \end{aligned}$$

<u>Proof of (2)</u>. This follows from (5.5.4).

<u>Proof of (3)</u>. Set

(5.8.4) $\delta_{\nu,\sigma}(\mu) = \gamma_{\nu,\sigma}(1 - \mu(1-\nu))$;

then $\delta_{\nu,\sigma}$ reparametrizes $\gamma_{\nu,\sigma}$ as a weak geodesic from $\gamma_{\nu,\sigma}(1) = \beta(\sigma)$ to $\gamma_{\nu,\sigma}(\nu) = \gamma_X(\nu)$, with domain $[0,1]$. For fixed ν and σ, $\partial_\nu \underline{\Delta}(\nu,\lambda,\sigma)$ is a Jacobi field along $\gamma_{\nu,\sigma}$ which vanishes at $\gamma_{\nu,\sigma}(1)$. Thus $\partial_\nu \underline{\Delta}(\nu,\lambda,\sigma)$ may be regarded as a Jacobi field along $\delta_{\nu,\sigma}$ from $\delta_{\nu,\sigma}(0) = \beta(\sigma)$. Since $L(\delta_{\nu,\sigma}) = L(\gamma_{\nu,\sigma}) < \underline{\tau}^\#$ by (5.5.3), I may apply Proposition 4.12:

$$\begin{aligned} (5.8.5)\quad |\partial_\nu \underline{\Delta}(\nu,\lambda,\sigma)| &\leq (3/2)((1-\lambda)/(1-\nu))|\partial_\nu \underline{\Delta}(\nu,\nu,\sigma)| \\ &< 2(1-\lambda)|\partial_\nu \underline{\Delta}(\nu,\nu,\sigma)|, \end{aligned}$$

since $\nu < 1/4$ by (5.1.7).

To calculate $\partial_\nu \underline{\Delta}(\nu,\nu,\sigma) = \partial_\nu \gamma_{\nu,\sigma}(\nu)$, I differentiate (5.8.4) with respect to ν, at $\mu = 1$. Since $\delta_{\nu,\sigma}(1) = \gamma_X(\nu)$, this gives -- putting

$\lambda = 1 - \mu(1-\nu)$ --

$$\gamma_X{}'(\nu) = \underline{\partial_\nu}\gamma_{\nu,\sigma}(\nu) + \frac{\partial\lambda}{\partial\nu}\underline{\partial_\lambda}\gamma_{\nu,\sigma}(\nu)$$

$$= \underline{\partial_\nu \Delta}(\nu,\nu,\sigma) + \frac{\partial\lambda}{\partial\nu}\gamma_{\nu,\sigma}{}'(\nu).$$

Since $\frac{\partial\lambda}{\partial\nu} = \mu$, this reduces at $\mu = 1$ to

(5.8.6) $\underline{\partial_\nu \Delta}(\nu,\nu,\sigma) = \gamma_X{}'(\nu) - \gamma_{\nu,\sigma}{}'(\nu).$

Now by Proposition 5.2.1,

$$(1-\nu)\gamma_X{}'(\nu) = \tilde{\beta}_\nu(0);$$

and

$$(1-\nu)\gamma_{\nu,\sigma}{}'(\nu) = (1-\nu)\tilde{\gamma}_{\nu,\sigma}{}'(\nu), \text{ by the exponential property of } e_N$$

$$= \tilde{\beta}_\nu(\sigma), \qquad \text{by (5.3.1).}$$

Hence

$$|\gamma_X{}'(\nu) - \gamma_{\nu,\sigma}{}'(\nu)| = (1-\nu)^{-1}|\tilde{\beta}_\nu(0) - \tilde{\beta}_\nu(\sigma)|$$

$$\leq (1-\nu)^{-1}L(\tilde{\beta}_\nu \restriction [0,\sigma])$$

$$\leq (4/3)\int_0^\sigma |\tilde{\beta}_\nu{}'(\hat{\sigma})|d\hat{\sigma}, \qquad \text{by (5.1.7)}$$

$$\leq 2B\sigma, \qquad \text{by (5.4.2).}$$

Substituting into (5.8.6) gives

(5.8.7) $|\underline{\partial_\nu \Delta}(\nu,\nu,\sigma)| \leq 2B\sigma.$

Now (3) follows from (5.8.5) and (5.8.7). q.e.d.

COROLLARY 5.8.8. For every $(\lambda,\sigma) \in (0,1] \times [0,1]$,

$$\lim_{\substack{\nu<\lambda \\ \nu\to 0^+}} \underline{\Delta}(\nu,\lambda,\sigma) \text{ exists.}$$

Moreover the function Δ_0: $[0,1] \times [0,1] \to M$ defined by

$$\Delta_0(\lambda,\sigma) = \begin{cases} P & \text{if } \lambda = 0 \\ \lim\limits_{\substack{\nu<\lambda \\ \nu\to 0^+}} \underline{\Delta}(\nu,\lambda,\sigma), & \text{if } \lambda > 0 \end{cases}$$

is continuous.

Proof. The existence and continuity of Δ_0 on $(0,1] \times [0,1]$ follows from Lemma 5.8.3, which shows that $\underline{\Delta}(\nu,\lambda,\sigma)$ converges as $\nu \to 0^+$ uniformly in λ and σ. To verify the continuity of Δ_0 at P,

$$(5.8.9)\quad \begin{aligned} d^M(P,\underline{\Delta}(\nu,\lambda,\sigma)) &= d^M(P,\gamma_{\nu,\sigma}(\lambda)) \\ &\leq L(\gamma_X \restriction [0,\nu]) + L(\gamma_{\nu,\sigma} \restriction [\nu,\lambda]) \\ &\leq \nu\tau(X) + ((\lambda-\nu)/(1-\nu))\underline{\tau}^{\#}, && \text{by Lemma 5.8.1} \\ &\leq \nu\tau(X) + \lambda\underline{\tau}^{\#}, && \text{by (1.17.4).} \end{aligned}$$

Let $\nu \to 0^+$; then

$$d^M(P,\Delta_0(\lambda,\sigma)) \leq \lambda\underline{\tau}^{\#},$$

for $(\lambda,\sigma) \in (0,1] \times [0,1]$. This proves that Δ_0 is continuous at $\lambda = 0$.

q.e.d.

PROPOSITION 5.9. Given $\lambda \in (0,1]$ let $\nu \in (0, \min\{1/4,\lambda\})$. Then

$$\Delta_0(\lambda,\sigma) = \lim_{\nu\to 0^+} \gamma_{\nu,\sigma}(\lambda) = \gamma_{\beta\sigma}(\lambda),$$

where $\gamma_{\beta\sigma}$ is as in (3.4). Moreover the convergence is uniform in λ and σ.

Proof. Let a decreasing sequence $\nu(n) \in (0,1/4)$ be given such that $\nu(n) \to 0$. For fixed σ, the family

$$\{\gamma_{\nu(n),\sigma}\colon\ [\nu(n),1] \to M^O\}$$

is umiformly bounded, since

$$(5.9.1)\quad d^M(P,\gamma_{\nu(n),\sigma}\lambda) \leq \nu(n)\tau(X) + \lambda\underline{\tau}^{\#}, \qquad \text{by (5.8.9)}$$

$$\leq (1/4)\tau(X) + \lambda\underline{\tau}^{\#}.$$

Also $\{\gamma_{\nu(n),\sigma}\}$ is equicontinuous by Lemma 5.8.1. By the Ascoli-Arzela theorem there is a subsequence $\nu(n^*)$ such that for each n_0^*, the sequence $\{\gamma_{\nu(n^*),\sigma}$ for $n^* \geq n_0^*\}$ converges uniformly on $[\nu(n_0^*),1]$. The limit path, $\gamma_{*,\sigma}$, say, is defined on $(0,1]$ and is continuous. From (5.9.1), taking the limit as $n^* \to \infty$,

$$d^M(P,\gamma_{*,\sigma}\lambda) \leq \lambda\underline{\tau}^{\#},$$

for all $\lambda \in (0,1]$. Thus $\lim_{\lambda\to 0^+} \gamma_{*,\sigma}(\lambda) = P$. Set $\gamma_{*,\sigma}(0) = P$; then $\gamma_{*,\sigma}$ is defined on $[0,1]$ and is now a continuous path starting from P. The next step is to show that $\gamma_{*,\sigma}$ is a path from P in the sense of (2.2) and that $\gamma_{*,\sigma}\restriction(0,1]$ is a weak geodesic in M^o.

Let $\lambda \in (0,1]$; I must show that $\gamma_{*,\sigma}(\lambda) \neq P$. Now

$$(5.9.2)\quad \begin{aligned} d^M(\gamma_X\lambda,\gamma_{\nu,\sigma}\lambda) &= d^M(\gamma_{\nu,0}\lambda,\gamma_{\nu,\sigma}\lambda) \\ &\leq L(\beta_\nu^\lambda\restriction[0,\sigma]) \\ &< (1/4)\lambda\tau(X), \end{aligned}$$

by (5.5.8). On the other hand,

$$d^M(P,\gamma_X\lambda) = \lambda\tau(X),$$

by (5.1.9). Therefore

$$\begin{aligned} d^M(P,\gamma_{*,\sigma}\lambda) &\geq \lambda\tau(X) - (1/4)\lambda\tau(X) \\ &= (3/4)\lambda\tau(X) \\ &> 0, \end{aligned}$$

since $\lambda > 0$. That is, $\gamma_{*,\sigma}(\lambda) \neq P$ whenever $\lambda \in (0,1]$. To show that $\gamma_{*,\sigma}$ is a path from P it remains to prove (2.2.1) -- (2.2.3); but these will follow immediately from showing that $\gamma_{*,\sigma}\restriction(0,1]$ is a weak geodesic, which I now do.

Let $\lambda \in (0,1]$ be fixed. Let $r > 0$ be such that (1.15.8) holds for $B(\gamma_{*,\sigma}\lambda,M;r)$ and that this ball is contained in M^o. Let $[\lambda',\lambda'']$ be a neighbourhood of λ in $(0,1]$ of length

$$\lambda'' - \lambda' < r/(4\underline{\tau}^{\#}).$$

There exists n_0^* such that whenever $n^* \geq n_0^*$, $\gamma_{\nu(n^*),\sigma}$ is defined on $[\lambda',\lambda'']$, and such that

$$d^M(\gamma_{\nu(n^*),\sigma}\lambda,\ \gamma_{*,\sigma}\lambda) < r/3.$$

By Lemma 5.8.1 and (5.1.7),

$$|\gamma_{\nu(n^*),\sigma}'| < (4/3)\underline{\tau}^{\#}$$

Hence

$$L(\gamma_{\nu(n^*),\sigma}\restriction[\lambda',\lambda'']) < (\lambda'' - \lambda')(4/3)\underline{\tau}^{\#}$$
$$< r/3.$$

It now follows that whenever $\mu \in [\lambda',\lambda'']$,

(5.9.3) $$d^M(\gamma_{*,\sigma}\lambda,\gamma_{\nu(n^*),\sigma}) \leq d^M(\gamma_{*,\sigma}\lambda,\gamma_{\nu(n^*),\sigma}\lambda) + L(\gamma_{\nu(n^*),\sigma}\restriction[\lambda',\lambda''])$$
$$< r/3 + r/3$$
$$= 2r/3.$$

Let $n^* \to \infty$; then

(5.9.4) $$d^M(\gamma_{*,\sigma}\lambda,\ \gamma_{*,\sigma}\mu) \leq 2r/3.$$

(5.9.3) and (5.9.4) show that, when restricted to $[\lambda',\lambda'']$, $\{\gamma_{\nu(n^*),\sigma}$ for $n^* \geq n_0^*\}$ and $\gamma_{*,\sigma}$ all have images in $B(\gamma_{*,\sigma}\lambda,M;r)$. Now each $\gamma_{\nu(n^*),\sigma}\restriction[\lambda',\lambda'']$ is a weak geodesic; so by (1.15.8) it is in fact a geodesic. By Lemma 2.10, the limit path, $\gamma_{*,\sigma}\restriction[\lambda',\lambda'']$, is also a geodesic. So by Proposition 2.9.3, $\gamma_{*,\sigma}\restriction(0,1]$ is locally a weak geodesic, and thus is overall a weak geodesic.

I have shown so far that $\gamma_{*,\sigma}$ is a weak geodesic from P to $\beta(\sigma)$; to complete the proof of Proposition 5.9 it remains to show that $\gamma_{*,\sigma}$ equals $\gamma_{\beta\sigma}$, which will be done by applying Proposition 3.12. Now

$$\begin{aligned} d^M(\gamma_X 1, \gamma_{\beta\sigma} 1) &= d^M(X, \beta\sigma) \\ &\leq L(\beta) \\ &< \underline{r}\tau(X), \end{aligned}$$

by (5.1.5). By (5.1.2) and (5.1.3), $\tau(X) < \underline{\tau}^{\#}$; hence (3.13.2) gives, for $\lambda \in (0,1]$,

$$\begin{aligned} d^M(\gamma_X\lambda, \gamma_{\beta\sigma}\lambda) &< 2\lambda H(\underline{r}\tau(X) + \underline{r}\tau(X))/h \\ &< (1/2)\lambda\underline{r}^*\tau(X), \end{aligned}$$

by (3.13.1). Combining this with (5.9.2) gives

$$\begin{aligned} d^M(\gamma_{\beta\sigma}\lambda, \gamma_{*,\sigma}\lambda) &< (1/2)\lambda\underline{r}^*\tau(X) + (1/4)\lambda\underline{r}^*\tau(X) \\ &< \lambda\underline{r}^*\tau(X), \end{aligned}$$

for all $\lambda \in (0,1]$. By Proposition 3.12, applied to the geodesic $\gamma_{\beta\sigma}$ and to the weak geodesic $\gamma_{*,\sigma}$ from P to $\beta(\sigma)$, $\gamma_{*,\sigma} = \gamma_{\beta\sigma}$. The first assertion of the proposition now follows, and the second is quoted from the proof of Corollary 5.8.8. q.e.d.

LEMMA 5.10. For $\lambda > 0$, $\underline{\partial_\sigma}\Delta_0(\lambda,0)$ exists and equals

$$\lim_{\substack{\nu<\lambda \\ \nu\to 0^+}} \underline{\partial_\sigma}\underline{\Delta}(\nu,\lambda,0).$$

<u>Proof</u>. λ will be fixed throughout this proof. For any $\nu < \lambda$, $\underline{\Delta}(\nu,\lambda,0) = \gamma_X(\lambda)$; hence

$$\text{(5.10.1)} \quad \begin{aligned} d^M(\gamma_X\lambda, \underline{\Delta}(\nu,\lambda,\sigma)) &\leq \int_0^\sigma |\underline{\partial_\sigma}\underline{\Delta}(\nu,\lambda,\sigma)|d\sigma \\ &\leq (3/2)\lambda\sigma B, \end{aligned}$$

by Lemma 5.8.2 and (5.1.6). Let $\phi: \mathbb{R}^n \to M^o$ be a C^∞ chart about $\gamma_X(\lambda)$. Then (5.10.1) implies that whenever σ is small enough, independently of ν, then $\underline{\Delta}(\nu,\lambda,\sigma) \in \mathrm{im}(\phi)$. Assume henceforth this restriction on σ. To simplify the notation, I shall suppress mention of ϕ and ϕ^{-1}, and regard M^o itself as being a vector space near $\gamma_X(\lambda) = 0$. Since $\|D\phi\|$ and $\|D\phi^{-1}\|$ may be assumed to be bounded, (5.10.1) may be expressed as

(5.10.2) $|\underline{\Delta}(\nu,\lambda,\sigma)| = O(\sigma;\ \text{fixed } \underline{\Delta} \text{ and } \phi)$.

Now let $0 < \nu^* < \nu$. Then

$$d^M(\underline{\Delta}(\nu^*,\lambda,\sigma),\ \underline{\Delta}(\nu,\lambda,\sigma)) \leq \int_{\nu^*}^{\nu} |\underline{\partial_\nu \Delta}(\hat{\nu},\lambda,\sigma)|\,d\hat{\nu}$$

$$\leq 4(1-\lambda)B\sigma(\nu-\nu^*),$$

by Lemma 5.8.3. Hence, applying ϕ^{-1},

(5.10.3) $|\underline{\Delta}(\nu^*,\lambda,\sigma) - \underline{\Delta}(\nu,\lambda,\sigma)| = O(\sigma(\nu-\nu^*);\ \text{fixed } \underline{\Delta} \text{ and } \phi)$.

In view of Lemma 5.8.2,

$$\lim_{\sigma\to 0^+} \underline{\Delta}(\nu,\lambda,\sigma)/\sigma \text{ exists and equals } \underline{\partial_\sigma \Delta}(\nu,\lambda,0);$$

and similarly for $\underline{\Delta}(\nu^*,\lambda,\sigma)$. Dividing (5.10.3) through by σ and taking the limit as $\sigma \to 0^+$ therefore gives

$$|\underline{\partial_\sigma \Delta}(\nu^*,\lambda,0) - \underline{\partial_\sigma \Delta}(\nu,\lambda,0)| = O(\nu-\nu^*;\ \text{fixed } \underline{\Delta} \text{ and } \phi).$$

It follows that

(5.10.4) $\psi(\lambda,0) = \lim_{\substack{\nu<\lambda \\ \nu\to 0^+}} \underline{\partial_\sigma \Delta}(\nu,\lambda,0)$ exists.

Since $\Delta_0(\lambda,0) = \gamma_X(\lambda) = \phi(0)$, it remains to show that given $\varepsilon > 0$, under ϕ^{-1},

(5.10.5) $|\Delta_0(\lambda,\sigma) - \sigma\psi(\lambda,0)| < \varepsilon\sigma$ whenever σ is sufficiently small.

Let $\nu^* \to 0^+$ in (5.10.3); this gives

$$|\Delta_0(\lambda,\sigma) - \underline{\Delta}(\nu,\lambda,\sigma)| = O(\sigma\nu;\ \text{fixed } \underline{\Delta} \text{ and } \phi)$$

$$< (\varepsilon/4)\sigma,$$

once ν is sufficiently small. I assume ν chosen so small that also, in (5.10.4),

$$|\psi(\lambda,0) - \underline{\partial_\sigma}\underline{\Delta}(\nu,\lambda,0)| < \varepsilon/4.$$

Fix such a ν; then the left-hand side of (5.10.5) is within $(\varepsilon/2)\sigma$ of

(5.10.6) $|\underline{\Delta}(\nu,\lambda,\sigma) - \sigma\underline{\partial_\sigma}\underline{\Delta}(\nu,\lambda,0)|$,

for all σ. But again, since $\underline{\Delta}(\nu,\lambda,0) = 0$ in the chart ϕ, (5.10.6) is $<(\varepsilon/2)\sigma$ whenever σ is sufficiently small, by definition of $\underline{\partial_\sigma}\underline{\Delta}(\nu,\lambda,0)$. Now (5.10.5) follows.

q.e.d.

LEMMA 5.11. $\lim_{\nu\to 0^+} \underline{\partial_\lambda}\underline{\Delta}(\nu,1,\sigma)$ exists and equals

$$\underline{\partial_\lambda}\Delta_0(1,\sigma) = \gamma_{\beta\sigma}'(1).$$

<u>Proof</u>. σ will be fixed throughout this proof. Since each $\gamma_{\nu,\sigma}$ is a weak geodesic, there is a Jacobi field along $\gamma_{\nu,\sigma}$ defined by $\underline{\partial_\nu}\underline{\Delta}(\nu,\lambda,\sigma)$. Its value at $\gamma_{\nu,\sigma}(1) = \beta(\sigma)$ is 0. By Corollary 4.12.10,

$$\begin{aligned}|D^M_{\gamma_{\nu,\sigma}'1}\, \underline{\partial_\nu}\underline{\Delta}|(\nu,1,\sigma) &\le (3/2)(1-\nu)^{-1}|\underline{\partial_\nu}\underline{\Delta}(\nu,\nu,\sigma)| \\ &\le (3/2)(4/3)2B\sigma, \qquad \text{by (5.1.7) and (5.8.7)} \\ &= 4B\sigma.\end{aligned}$$

The order of taking partial derivatives may be reversed; hence

$$|D^M_{\underline{\partial_\nu}}\, \underline{\partial_\lambda}\underline{\Delta}|(\nu,1,\sigma) \le 4B\sigma.$$

In particular this derivative is bounded. Therefore

(5.11.1) $\psi = \lim_{\nu\to 0^+} \underline{\partial_\lambda}\underline{\Delta}(\nu,1,\sigma)$ exists,

and for $0 < \nu^* < \nu$,

(5.11.2) $|\underline{\partial_\lambda}\underline{\Delta}(\nu^*,1,\sigma) - \underline{\partial_\lambda}\underline{\Delta}(\nu,1,\sigma)| \le 4B\sigma(\nu-\nu^*)$.

Let γ_ψ be the geodesic in M^0 such that $\gamma_\psi(1) = \beta(\sigma)$ and $\gamma_\psi'(1) = \psi$, defined on some neighbourhood of 1 in [0,1]. Now $\gamma_{\nu,\sigma}$ has final conditions $\gamma_{\nu,\sigma}(1) = \beta(\sigma)$ and $\gamma_{\nu,\sigma}'(1) = \underline{\partial_\lambda}\underline{\Delta}(\nu,1,\sigma)$, defined on $[\nu,1] \supseteq [1/4,1]$.

Hence on some common domain, in view of (5.11.1),

$$\gamma_\psi = \lim_{\nu\to 0^+} \gamma_{\nu,\sigma}$$
$$= \gamma_{\beta\sigma},$$

by Proposition 5.9. Therefore

$$\psi = \gamma_\psi'(1) = \gamma_{\beta\sigma}'(1),$$

and the lemma follows.

Observe for future reference

(5.11.3) $|\underline{\partial}_\lambda \underline{\Delta}(\nu,1,\sigma) - \underline{\partial}_\lambda \Delta_0(1,\sigma)| \leq 4B\sigma\nu,$

which follows from (5.11.2) by taking the limit as $\nu^* \to 0^+$. q.e.d.

Relationship with Jacobi Fields

(5.12) For the rest of this chapter I shall assume that

(5.12.1) β is a C^∞ path in $B^O(P,M;\hat{\underline{\tau}})$,

unless otherwise stated, where $\hat{\underline{\tau}}$ is as in (5.1.2). For example, if $X \in B^O(P,M;\hat{\underline{\tau}}/2)$ and $Y \in B(X,M;\tau(X))$, then any geodesic β from X to Y satisfies (5.12.1).

Let β satisfy (5.12.1). Then for each σ, there is a unique geodesic $\gamma_{\beta\sigma}$ from P to $\beta(\sigma)$, and there is a unique Jacobi field

(5.12.2) $\underline{J}\{\sigma\}$ along $\gamma_{\beta\sigma}$ from P with final condition $\underline{J}\{\sigma\}(1) = \beta'(\sigma)$.
($\underline{J}\{\sigma\}$ is an abbreviation for $\underline{J}\{\beta'\sigma\}$ of (4.14).)

THEOREM 5.13. Let β satisfy (5.12.1). For each σ let $\underline{J}\{\sigma\}$ be as in (5.12.2). Then

$$\underline{\partial}_\sigma \gamma_{\beta\sigma}(\lambda) \text{ exists and equals } \underline{J}\{\sigma\}(\lambda),$$

for all $\lambda \in (0,1]$.

Proof. Two preliminary simplifications can be made.

(5.13.1) It is sufficient to consider the case $\sigma = 0$.

(5.13.2) I may assume that $L(\beta) < \underline{r}\tau(X)$, where $X = \beta(0)$; that is, that β satisfies (5.1.5). For otherwise β can be replaced by a C^∞ path β^* in $B^o(P,M;\hat{\underline{\tau}})$ such that β^* satisfies (5.1.5) and equals β on some initial segment.

Let $\underline{\Delta}$ be as in (5.7.1) and Δ_0 as in Corollary 5.8.8. By Proposition 5.9, $\Delta_0(\lambda,\sigma) = \gamma_{\beta\sigma}(\lambda)$. Hence

$$\underline{\partial}_\sigma\gamma_{\beta\sigma}(\lambda)|_{\sigma=0} = \underline{\partial}_\sigma\Delta_0(\lambda,0)$$

$$= \lim_{\substack{\nu<\lambda \\ \nu\to 0^+}} \underline{\partial}_\sigma\underline{\Delta}(\nu,\lambda,0), \quad \text{by Lemma 5.10}$$

$$= \lim_{\substack{\nu<\lambda \\ \nu\to 0^+}} \underline{\partial}_\sigma\gamma_{\nu,\sigma}(\lambda)|_{\sigma=0},$$

by (5.7.1) and (5.3.2). Let $\underline{J}_\nu$ be the Jacobi field along $\gamma_X\restriction[\nu,1]$ with boundary conditions $\underline{J}_\nu(\nu) = 0$, $\underline{J}_\nu(1) = \beta'(0)$ (so $\underline{J}_\nu$ satisfies (4.11.3)). Since each $\gamma_{\nu,\sigma}$ is a weak geodesic, $\underline{\partial}_\sigma\gamma_{\nu,\sigma}|_{\sigma=0}$ is a Jacobi field along $\gamma_{\nu,0}\restriction[\nu,1] = \gamma_X\restriction[\nu,1]$. Now $\underline{\partial}_\sigma\gamma_{\nu,\sigma}|_{\sigma=0}$ has the same boundary conditions as does $\underline{J}_\nu$. It follows from Corollary 4.12.9 (compare Hicks [8, p. 147]) that

(5.13.3) $\underline{\partial}_\sigma\gamma_{\nu,\sigma}|_{\sigma=0} = \underline{J}_\nu$.

Thus

$$\underline{\partial}_\sigma\gamma_{\beta\sigma}(\lambda)|_{\sigma=0} = \lim_{\substack{\nu<\lambda \\ \nu\to 0^+}} \underline{J}_\nu(\lambda)$$

$$= \underline{J}\{0\}(\lambda),$$

by Theorem 4.13.1. In view of (5.13.1), this proves the theorem.

q.e.d.

LEMMA 5.14. Let $X \in B^o(P,M;\hat{\underline{\tau}}/2)$ and let $Y \in B(X,M;\tau(X))$. Then

$$d^M(\gamma_X\lambda,\gamma_Y\lambda) \le (3/2)\lambda d^M(X,Y),$$

for all $\lambda \in [0,1]$.

<u>Proof</u>. Let β be a geodesic from X to Y; then $\mathrm{im}(\beta) \subseteq B^o(P,M;\hat{\underline{\tau}})$ and β is C^∞

by Proposition 2.13. Set

$$\beta^\lambda(\sigma) = \gamma_{\beta\sigma}(\lambda).$$

By Theorem 5.13, β^λ is differentiable and

$$\beta^{\lambda\prime}(\sigma) = \underline{J}\{\sigma\}(\lambda),$$

where $\underline{J}\{\sigma\}$ is as in (5.13.2). By Corollary 4.15.7,

$$(5.14.1) \quad \begin{aligned} |\beta^{\lambda\prime}(\sigma)| &= |\underline{J}_\nu\{\sigma\}(\lambda)| \\ &\leq (3/2)\lambda|\underline{J}\{\sigma\}(1)| \\ &= (3/2)\lambda|\beta'(\sigma)|. \end{aligned}$$

Now

$$\begin{aligned} d^M(\gamma_X\lambda,\gamma_Y\lambda) &\leq L^*(\beta^\lambda), \quad \text{by (2.5.6)} \\ &\leq (3/2)\lambda\int_0^1 |\beta'(\sigma)|d\sigma, \text{ by (5.14.1) and Lemma 2.5} \\ &= (3/2)\lambda L(\beta) \\ &= (3/2)\lambda d^M(X,Y). \end{aligned}$$

q.e.d.

PROPOSITION 5.15. Let β: $[0,1] \to B^o(P,M;\hat{\underline{\tau}}/2)$ be a differentiable path (not necessarily C^∞). Then

$$\underline{\partial}_\sigma\gamma_{\beta\sigma}(\lambda) \text{ exists and equals } \underline{J}\{\sigma\}(\lambda)$$

for all $\lambda \in (0,1]$.

Proof. It is sufficient to prove the proposition under the simplifying assumptions (5.13.1) and (5.13.2). Let β^*: $[0,1] \to M^o$ be a C^∞ path satisfying (5.1.5) and such that $\beta^{*\prime}(0) = \beta'(0)$. Then

$$(5.15.1) \quad d^M(\beta^*\sigma,\beta\sigma) = o(\sigma;\ \text{fixed } \beta \text{ and } \beta^*).$$

A cruder estimate is

$$(5.15.2) \quad d^M(\beta^*\sigma,\beta\sigma) \leq L(\beta^*) + L(\beta)$$

$$\leq 2\underline{r}\tau(X), \quad \text{by (5.1.5) and (5.13.2)}$$
$$< \tau(X),$$

since $\underline{r} < 1/16$ by (5.1). Therefore Lemma 5.14 applies to give

$$\text{(5.15.3)} \quad d^M(\gamma_{\beta*\sigma}\lambda, \gamma_{\beta\sigma}\lambda) \leq (3/2)\lambda d^M(\beta*\sigma, \beta\sigma)$$
$$= o(\sigma;\ \text{fixed } \beta \text{ and } \beta*),$$

by (5.15.1). By Theorem 5.13, $\underline{\partial}_\sigma\gamma_{\beta*\sigma}(\lambda)|_{\sigma=0}$ exists and equals

$$\underline{J}\{\beta*'0\}(\lambda) = \underline{J}\{\beta'0\}(\lambda) = \underline{J}\{0\}(\lambda).$$

The proof of Lemma 5.10, together with (5.15.3), shows that $\underline{\partial}_\sigma\gamma_{\beta\sigma}(\lambda)|_{\sigma=0}$ exists and equals $\underline{\partial}_\sigma\gamma_{\beta*\sigma}(\lambda)|_{\sigma=0}$. The proposition now follows. q.e.d.

LEMMA 5.16. Let β satisfy (5.12.1) and let $\underline{\Gamma}$ be the vector field of (3.4.3). Then $\underline{\Gamma}(\beta\sigma) = \gamma_{\beta\sigma}'(1)$ is differentiable along β, and

$$D^M_{\beta'}\underline{\Gamma} = D^M_{\gamma_{\beta\sigma}'1}\ \underline{J}\{\sigma\}.$$

<u>Proof</u>. It is sufficient to consider the case $\sigma = 0$, as in (5.13.1). Let $\underline{J}_\nu$ be the Jacobi field along $\gamma_X \restriction [\nu,1]$ satisfying (4.11.3) with final condition $\beta'(0)$. By Lemma 4.16, $D^M_{\gamma_{\beta 0}'1}\ \underline{J}\{0\}$ exists and equals

$$\text{(5.16.1)} \quad \lim_{\nu\to 0^+} D^M_{\gamma_{\beta 0}'1}\ \underline{J}_\nu = \lim_{\nu\to 0^+} D^M_{\gamma_{\beta 0}'1}\ \underline{\partial}_\sigma\gamma_{\nu,\sigma}|_{\sigma=0}, \quad \text{by (5.13.3)}$$
$$= \lim_{\nu\to 0^+} (D^M_{\gamma_{\beta 0}'1}\ \underline{\partial}_\sigma\underline{\Delta})(\nu,1,0), \quad \text{by (5.7.1) and (5.3.2)}$$
$$= \lim_{\nu\to 0^+} (D^M{}_{'0}\ \underline{\partial}_\lambda\underline{\Delta})(\nu,1,0),$$

since the order of taking partial derivatives may be reversed. Let $\phi\colon \mathbb{R}^n \to M^o$ be a chart about $\beta(0)$ which uses Fermi coordinates along β (see Hicks [8, p. 133]). To simplify the notation I suppress mention of ϕ and ϕ^{-1}, as in the proof of Lemma 5.10. Then $D^M_{\beta'\sigma}$ becomes simply the partial derivative $\underline{\partial}_\sigma$, applied at $\beta(\sigma)$. It is therefore required to prove that given $\varepsilon > 0$,

(5.16.2) $\quad |\underline{\Gamma}(\beta\sigma) - \underline{\Gamma}(\beta 0) - \sigma D^M_{\gamma_{\beta\sigma}{}',1}\, \underline{J}\{0\}| < \varepsilon\sigma,$

whenever σ is sufficiently small. Now

$$\begin{aligned}\underline{\Gamma}(\beta\sigma) &= \gamma_{\beta\sigma}{}'(1)\\ &= \underline{\partial}_\lambda \Delta_0(1,\sigma),\end{aligned}$$

by Proposition 5.9. So by (5.11.3),

$$\begin{aligned}|\underline{\Gamma}(\beta\sigma) - \underline{\partial}_\lambda\underline{\Delta}(\nu,1,\sigma)| &\le 4B\sigma\nu\\ &< (\varepsilon/4)\sigma,\end{aligned}$$

once ν is chosen small enough. Let ν be fixed so small that also in (5.16.1),

$$|D^M_{\gamma_{\beta 0}{}',1}\, \underline{J}\{0\} - D^M_{\beta',0}\, \underline{\partial}_\lambda\underline{\Delta}(\nu,1,0)| < \varepsilon/4.$$

Since

$$\underline{\Gamma}(\beta 0) = \gamma_{\beta 0}{}'(1) = \underline{\partial}_\lambda\underline{\Delta}(\nu,1,0),$$

it follows that the left-hand side of (5.16.2) is within $(\varepsilon/2)\sigma$ of

(5.16.3) $\quad |\underline{\partial}_\lambda\underline{\Delta}(\nu,1,\sigma) - \underline{\partial}_\lambda\underline{\Delta}(\nu,1,0) - \sigma D^M_{\beta',0}\underline{\partial}_\lambda\underline{\Delta}(\nu,1,0)|.$

In this context, under ϕ^{-1},

$$D^M_{\beta',0}\underline{\partial}_\lambda\underline{\Delta} = \underline{\partial}_\sigma\underline{\partial}_\lambda\underline{\Delta};$$

and since $\underline{\Delta}$ is C^∞, (5.16.3) is $<(\varepsilon/2)\sigma$ whenever σ is sufficiently small. Now (5.16.2) follows, and with it, Lemma 5.16. q.e.d.

PROPOSITION 5.17. Given $\underline{\lambda} \in (0,1]$ and a C^∞ path $\beta^{\underline{\lambda}}$ in $B^o(P,M;\underline{\lambda\hat{\tau}}/2)$, let $\underline{J}^{\underline{\lambda}}\{\sigma\}$ be the Jacobi field along $\gamma_{\beta^{\underline{\lambda}}\sigma}$ from P with final condition $\beta^{\underline{\lambda}}{}'(\sigma)$, for every $\sigma \in [0,1]$. Then:

(1) There exists a differentiable path β in $B^o(P,M;\underline{\hat{\tau}}/2)$ such that, for all σ,

$$\gamma_{\beta\sigma}(\underline{\lambda}) = \beta^{\underline{\lambda}}(\sigma);$$

(2) $\underline{J}\{\sigma\}(\underline{\lambda}\mu) = \underline{J}^{\underline{\lambda}}\{\sigma\}(\mu)$, for all $\mu \in (0,1]$,

where $\underline{J}\{\sigma\}$ is as in (5.12.2).

Proof of (1). Let γ_σ be the geodesic from P to $\beta^{\underline{\lambda}}(\sigma)$ with domain $[0,\underline{\lambda}]$. Then $|\gamma_\sigma'| < \hat{\underline{\tau}}/2 < \underline{\tau}^{\#}$, by (5.1.2). Hence by Lemma 3.17, γ_σ can be extended to a geodesic with domain [0,1]. Since there can be only one geodesic in the Riemannian manifold M^{O} with domain (0,1] and given initial segment on $(0,\underline{\lambda}]$, it follows that γ_σ is unique; and β must therefore be defined by $\beta(\sigma) = \gamma_\sigma(1)$. Now $\gamma_{\beta\sigma} = \gamma_\sigma$, so

$$\begin{aligned}\gamma_{\beta\sigma}(\underline{\lambda}) &= \gamma_\sigma(\underline{\lambda})\\ &= \beta^{\underline{\lambda}}(\sigma)\end{aligned}$$

as required. It remains to show that β is differentiable. First observe that for $\mu \in [0,1]$,

$$\begin{aligned}(5.17.3)\quad \gamma_{\beta\sigma}(\underline{\lambda}\mu) &= \gamma_\sigma(\underline{\lambda}\mu)\\ &= \gamma_{\beta^{\underline{\lambda}}\sigma}(\mu);\end{aligned}$$

and hence

$$\begin{aligned}(5.17.4)\quad \gamma_{\beta\sigma}'(\underline{\lambda}) &= \underline{\lambda}^{-1}\gamma_{\beta^{\underline{\lambda}}\sigma}'(1)\\ &= \underline{\lambda}^{-1}\underline{\Gamma}(\beta^{\underline{\lambda}}\sigma).\end{aligned}$$

Now a general weak geodesic γ in the Riemannian manifold M^{O} with domain $[\underline{\lambda},1]$ varies in C^∞ fashion with its three parameters: $\lambda \in [\underline{\lambda},1]$, $\gamma(\underline{\lambda})$ and $\gamma'(\underline{\lambda})$. In the case at hand, when $\gamma = \gamma_{\beta\sigma}$, the parameter $\gamma(\underline{\lambda}) = \beta^{\underline{\lambda}}(\sigma)$ is C^∞ in σ, and the parameter $\gamma'(\underline{\lambda})$ is differentiable with respect to σ by (5.17.4) and Lemma 5.16. It follows that $\beta(\sigma) = \gamma_{\beta\sigma}(1)$ is differentiable with respect to σ, which proves (1).

Proof of (2). This follows from (5.17.3) by taking the derivative in σ at $\sigma = 0$. q.e.d.

CORLLARY 5.17.5. Under the hypotheses of Proposition 5.17, set $\beta^{\mu}(\sigma) = \gamma_{\beta\sigma}(\mu)$ for $\mu \in (0,1]$. Then β^{μ} is differentiable.

LEMMA 5.18. Let the vector fields $\underline{\Gamma}$ and $\underline{\partial}_{\tau}$ on $B^{o}(P,M;\hat{\underline{\tau}})$ be as in (3.4.3) and (3.4.4). Given $X \in B^{o}(P,M;\hat{\underline{\tau}})$ and $\psi \in T_X M$, then:

(1) $D^{M}_{\psi}\underline{\Gamma}(X)$ exists and

$$|D^{M}_{\psi}\underline{\Gamma}(X) - \psi| = O(\tau(X)^{2k});$$

(2) $D^{M}_{\psi}\underline{\partial}_{\tau}(X)$ exists and

$$|D^{M}_{\psi}\underline{\partial}_{\tau}(X) - \tau(X)^{-1}\underline{\partial}_{\tau}{}^{\perp}(\psi)| = O(\tau(X)^{-1+2k})|\psi|.$$

(1.6.3,VII) is needed for this result.

<u>Proof of (1)</u>. Let β be a C^{∞} curve in $B^{o}(P,M;\hat{\underline{\tau}})$ such that

(5.18.3) $\beta(0) = X, \ \beta'(0) = \psi.$

Then by Lemma 5.16, and using its notation,

(5.18.4) $D^{M}_{\beta'0}\underline{\Gamma}(\beta 0) = D^{M}_{\gamma_X'1}\underline{J}\{0\}.$

In particular the right-hand side is independent of the choice of β satisfying (5.18.3), and may therefore be used to define $D^{M}_{\psi}\underline{\Gamma}(X)$. Now (1) follows from (5.18.4) and Lemma 4.17 (which requires (1.6.3,VII)).

<u>Proof of (2)</u>. By (3.4.4),

$$\underline{\partial}_{\tau}(X) = |\underline{\Gamma}(X)|^{-1}\underline{\Gamma}(X).$$

In view of (1), $|\underline{\Gamma}(X)|$ is differentiable in the direction ψ, and

$$\underline{\partial}_{\psi}|\underline{\Gamma}(X)| = |\underline{\Gamma}(X)|^{-1}\langle \underline{\Gamma}, D^{M}_{\psi}\underline{\Gamma}\rangle (X).$$

It follows that $D^{M}_{\psi}\underline{\partial}_{\tau}(X)$ exists, and moreover that

(5.18.5) $$D^{M}_{\psi}\underline{\partial}_{\tau}(X) = |\underline{\Gamma}(X)|^{-1}(D^{M}_{\psi}\underline{\Gamma})(X) - |\underline{\Gamma}(X)|^{-3}\langle \underline{\Gamma}, D^{M}_{\psi}\underline{\Gamma}\rangle (X)\underline{\Gamma}(X)$$

$$= |\underline{\Gamma}(X)|^{-1}\{D^M_\psi\underline{\Gamma}(X) - \langle \underline{\partial}_\tau, D^M_\psi\underline{\Gamma}\rangle(X)\underline{\partial}_\tau(X)\}$$

$$= |\underline{\Gamma}(X)|^{-1}\{D^M_\psi\underline{\Gamma}(X) - \underline{\partial}_\tau{}^{\parallel}(D^M_\psi\underline{\Gamma})(X)\}$$

$$= |\underline{\Gamma}(X)|^{-1}\underline{\partial}_\tau{}^{\perp}(D^M_\psi\underline{\Gamma})(X).$$

Set

$$\delta = D^M_\psi\underline{\Gamma}(X) - \psi.$$

Then, since $|\underline{\Gamma}(X)| = \tau(X)$, (5.18.5) becomes

$$D^M_\psi\underline{\partial}_\tau(X) = \tau(X)^{-1}\underline{\partial}_\tau{}^{\perp}(\psi+\delta)$$

Hence

$$\begin{aligned} |D^M_\psi\underline{\partial}_\tau(X) - \tau(X)^{-1}\underline{\partial}_\tau{}^{\perp}(\psi)| &= \tau(X)^{-1}|\partial_\tau{}^{\perp}(\delta)| \\ &\le \tau(X)^{-1}|\delta| \\ &= 0(\tau(X)^{-1+2k})|\psi|, \end{aligned}$$

by (1). q.e.d.

CHAPTER 6: THE EXPONENTIAL MAP AT P

Introduction

The first half of this chapter completes the proof of Theorem 1.7, that there exists an exponential map at P (Theorem 6.9). Properties (1.1) and (1.2) have already been shown (Corollary 2.17.4 and Theorem 3.3 respectively). The proof of (1.3), that e_P^{-1} is one-to-one, requires those parts of Chapters 4 and 5 that depend on (1.6.3,VII). To see why this surely must be so, let $X,Y \in M^o$ and let γ_X and γ_Y be the geodesics with domain [0,1] from P to X and Y respectively; it is required to show that $d^M(\gamma_X\lambda,\gamma_Y\lambda)/\lambda$ is bounded away from 0. Let β^λ be a geodesic from $\gamma_X(\lambda)$ to $\gamma_Y(\lambda)$, and to simplify the discussion, assume that $\mathrm{im}(\beta^\lambda) \subseteq M^o$. Then there exists a path β from X to Y such that for all σ, $\gamma_{\beta\sigma}(\lambda) = \beta^\lambda(\sigma)$, where $\gamma_{\beta\sigma}$ is the geodesic from P to $\beta(\sigma)$. We need an estimate of the form

$$L(\beta^\lambda) \geq (\text{constant})\cdot\lambda L(\beta);$$

then the left-hand side equals $d^M(\gamma_X\lambda,\gamma_Y\lambda)$ and the right-hand side is $\geq$ $(\text{constant})\cdot\lambda d^M(X,Y)$; so it will follow that $e_P^{-1}(X) \neq e_P^{-1}(Y)$. But such an estimate requires some sort of bound from below on the curvature of M, so (1.6.3,VII) seems to be indicated. The argument above is completed in Proposition 6.4.1. As a by-product, the continuity of e_P^{-1} and its inverse map e_P falls out. It remains to prove (1.4), that $\mathrm{im}(e_P^{-1})$ is a neighbourhood of P in T_PM. Let h be a chart at P; then h induces a homeomorphism $e(h)_P$: $T_PM \to M$ near P, by Remark 2.1. Then e_P^{-1} is close to $e(h)_P^{-1}$; in particular, near P these maps are homotopic through maps of pairs $(M,M^o) \to (T_PM,(T_PM)^o)$. Since M^o is homotopy-equivalent to a closed, compact manifold L^{n-1}, $e(h)_P^{-1}$ has a non-zero degree (mod 2); therefore so does e_P^{-1}; this implies that e_P^{-1} is surjective near P (Lemma 6.7).

It may be useful to note that (1.1), (1.2), (1.4) and half of (1.5) -- the continuity of e_P^{-1} -- are all proved without recourse to (1.6.3,VII).

The second half of the chapter analyses e_P: $T_PM \to M$ (near P).

Theorem 6.16 shows that the only thing that prevents e_P from being a chart at P in the sense of (1.6) is the lack of a C^∞ structure on the unit sphere

$$S_PM = \{\hat{X} \in T_PM \text{ such that } |\hat{X}| = 1\}$$

in T_PM. To show that such a structure exists is tantamount to showing that the unit radial vector field $\underline{\partial}_t$ on $(T_PM)^o$ (see (6.11.4)) is C^∞, or equivalently, that the corresponding field $\underline{\partial}_\tau$ on M^o is C^∞. Now Lemma 5.18.2 says that $\underline{\partial}_\tau$ can be differentiated at any $X \in M^o$ in any direction $\psi \in T_XM$; but this is short of proving $\underline{\partial}_\tau$ differentiable, let alone C^1, let alone C^∞. To make further progress in this direction requires, I believe, third- and higher-order differentiability assumptions about the Riemannian metric of M near P.

From the point of view of Remark 2.1, even if e_P were a chart near P, one would like it to be metrically the "best possible" chart. This ought to mean that S_PM should have a Riemannian metric ds^S of class C^∞, which would then induce a Riemannian metric ds^T on $(T_PM)^o$ by (1.16.6); and further, that d^T should be the intrinsic metric associated to ds^T as in (2.4.1). The correct candidate for the Riemannian metric ds^T is identified in (6.11.3). Unfortunately I cannot prove ds^T even continuous, so the "length" of a C^1 path in the sense of (2.3.1)does not make sense, and I do not see how (2.4.1) can be proved. Lemma 6.14 shows one satisfactory property of ds^T: it behaves correctly with respect to dilations, as in (1.16.8). Another positive result is Theorem 6.15, which says that d^T is an intrinsic metric in the sense of Aleksandroff (2.5.5).

The Existence of $\exp_P$

(6.1) Notation. The notation of (5.1) up to (5.1.4) will be used in this section, with the following addition to (5.1.3). Set

(6.1.1) $\underline{\tau} = \hat{\tau}/2;$

it will be assumed henceforth that

(6.1.2) $0 < \tau(X) < \underline{\tau},$

so that $X \in B^{o}(P,M;\underline{\tau})$. Let Y also denote a point of $B^{o}(P,M;\underline{\tau})$ and let γ_Y be as in (3.4). Then any geodesic β^{λ} from $\gamma_X(\lambda)$ to $\gamma_Y(\lambda)$ satisfies

(6.1.3) $\operatorname{im}(\beta^{\lambda}) \subseteq B(P,M;\lambda\hat{\underline{\tau}})$.

LEMMA 6.2. Whenever $0 \leq \mu \leq \lambda \leq 1$,

$$d^M(\gamma_X\mu,\gamma_Y\mu) = (\mu/\lambda)(1 + O((\lambda\underline{\tau})^{1/q}))d^M(\gamma_X\lambda,\gamma_Y\lambda),$$

where $1/2 \leq 1 + O((\lambda\underline{\tau})^{1/q}) \leq 3/2$.

Proof. Let β^{λ} be a C^{∞} path in $B^{o}(P,M;\lambda\hat{\underline{\tau}})$ from $\gamma_X(\lambda)$ to $\gamma_Y(\lambda)$. By Proposition 5.17.1 there exists a differentiable path β: $[0,1] \to B^{o}(P,M;\hat{\underline{\tau}})$ such that $\gamma_{\beta\sigma}(\lambda) = \beta^{\lambda}(\sigma)$. Moreover the path β^{μ}, defined by $\beta^{\mu}(\sigma) = \gamma_{\beta\sigma}(\mu)$, is also differentiable by Corollary 5.17.5. For each $\sigma \in [0,1]$ let $\underline{J}\{\sigma\}$ be as in (5.12.2). Then by Proposition 5.15,

$$\beta^{\lambda\prime}(\sigma) = \underline{J}\{\sigma\}(\lambda) \quad \text{and} \quad \beta^{\mu\prime}(\sigma) = \underline{J}\{\sigma\}(\mu).$$

In Corollary 4.19.5, I may take $T = \hat{\underline{\tau}}$; and since $\mu \leq \lambda$,

$$|\underline{J}\{\sigma\}(\mu)| = (\mu/\lambda)(1 + O((\lambda\hat{\underline{\tau}})^{1/q}))|\underline{J}\{\sigma\}(\lambda)|,$$

where

$$1/2 \leq 1 + O((\lambda\hat{\underline{\tau}})^{1/q}) \leq 3/2.$$

By (6.1.1),

$$O((\lambda\hat{\underline{\tau}})^{1/q}) = O((\lambda\underline{\tau})^{1/q});$$

so

$$|\beta^{\mu\prime}(\sigma)| = (\mu/\lambda)(1 + O((\lambda\underline{\tau})^{1/q}))|\beta^{\lambda\prime}(\sigma)|.$$

It follows from Lemma 2.5 that

(6.2.1) $$L^*(\beta^{\mu}) = (\mu/\lambda)(1 + O((\lambda\underline{\tau})^{1/q}))L(\beta^{\lambda}).$$

In view of (6.1.3), there are now two possibilities.

Case I. Assume there exists a geodesic β^λ from $\gamma_X(\lambda)$ to $\gamma_Y(\lambda)$ in $B^o(P,M;\hat{\underline{\tau}})$. Then by (6.2.1) and (2.5.6),

$$(6.2.2)\quad d^M(\gamma_X\mu,\gamma_Y\mu) \le L^*(\beta^\mu)$$
$$= (\mu/\lambda)(1 + O((\lambda\underline{\tau})^{1/q}))d^M(\gamma_X\lambda,\gamma_Y\lambda).$$

Case II. If Case I fails, then a geodesic β^λ from $\gamma_X(\lambda)$ to $\gamma_Y(\lambda)$ must go through P. By Theorem 3.3, β^λ consists of a reparametrization of the portions of γ_X from $\gamma_X(\lambda)$ to P and of γ_Y from P to $\gamma_Y(\tau)$. Hence

$$(6.2.3)\quad d^M(\gamma_X\mu,\gamma_Y\mu) \le d^M(\gamma_X\mu,P) + d^M(P,\gamma_Y\mu)$$
$$= \mu\tau(X) + \mu\tau(Y)$$
$$= (\mu/\lambda)(\lambda\tau(X) + \lambda\tau(Y))$$
$$= (\mu/\lambda)(d^M(\gamma_X\lambda,P) + d^M(P,\gamma_Y\lambda))$$
$$= (\mu/\lambda)d^M(\gamma_X\lambda,\gamma_Y\lambda).$$

(6.2.2) and (6.2.3) prove that the left-hand side of Lemma 6.2 is no greater than the right-hand side. The reverse inequality is derived similarly from (6.2.1) by interchanging the roles of λ and μ in (6.2.2) and in (6.2.3). q.e.d.

LEMMA 6.3. Let X and Y be as in (6.1). Then:

(1) $\lim_{\lambda\to 0^+} d^M(\gamma_X\lambda,\gamma_Y\lambda)/\lambda$ exists;

(2) $d^T(\hat{\gamma}_X,\hat{\gamma}_Y) = \lim_{\lambda\to 0^+} d^M(\gamma_X\lambda,\gamma_Y\lambda)/\lambda$;

(3) $(1/2)d^M(X,Y) \le d^T(\hat{\gamma}_X,\hat{\gamma}_Y) \le (3/2)d^M(X,Y)$.

Proof of (1). By Lemma 6.2, replacing λ by 1 and μ by λ,

$$(6.3.4)\quad d^M(\gamma_X\lambda,\gamma_Y\lambda) \le (3/2)\ d^M(X,Y).$$

Again by Lemma 6.2, whenever $\mu \le \lambda$,

$$|d^M(\gamma_X\mu,\gamma_Y\mu)/\mu - d^M(\gamma_X\lambda,\gamma_Y\lambda)/\lambda|$$

$$= O((\lambda\underline{\tau})^{1/q})d^M(\gamma_X\lambda,\gamma_Y\lambda)/\lambda$$

$$\leq (3/2)d^M(X,Y)O((\lambda\underline{\tau})^{1/q}),$$

by (6.3.4). Assertion (1) now follows.

Proof of (2). This follows from (1) and the definition of d^T in (2.25.1).

Proof of (3). This follows from Lemma 6.2 with $\lambda = 1$, by taking the limit as $\mu \to 0^+$ and applying (1) and (2). q.e.d.

COROLLARY 6.3.5. e_P^{-1} is continuous on $B(P,M;\underline{\tau})$; in fact it satisfies a uniform Lipschitz condition on this domain.

Proof. This follows immediately from the second inequality of Lemma 6.3.3. q.e.d.

PROPOSITION 6.4. (1) e_P^{-1} is one-to-one on $B(P,M;\underline{\tau})$;

(2) The inverse function

$$e_P\colon\ e_P^{-1}(B(P,M;\underline{\tau})) \to B(P,M;\underline{\tau})$$

is continuous; in fact it satisfies a uniform Lipschitz condition.

Proof of (1). Let $X,Y \in B(P,M;\underline{\tau})$ be such that $e_P^{-1}(X) = \hat{\gamma}_X$ equals $e_P^{-1}(Y) = \hat{\gamma}_Y$. If $X,Y \in M^o$, assertion (1) follows from the first inequality of Lemma 6.3.3. If X, say, equals P, the assertion follows from

$$\text{(6.4.3)}\quad d^M(P,Y) = \tau(Y)$$

$$= |\hat{\gamma}_Y|, \qquad \text{by (3.4.1)}$$

$$= d^T(\hat{P},\hat{\gamma}_Y), \quad \text{by (2.25.2)}.$$

Since $\hat{P} = e_P^{-1}(X)$, the last term is 0; so $Y = P$.

Proof of (2). If $X \in B^o(P,M;\underline{\tau})$, then the continuity of e_P at $\hat{\gamma}_X$ with Lipschitz constant 2 follows from the first inequality of Lemma 6.3.3. If $X = P$, then continuity at $\hat{P}$ with Lipschitz constant 2 follows from (6.4.3). q.e.d.

(6.5) Definition. The map e_P of Proposition 6.4.2 is called the exponential map of M at P.

Now let $h\colon cL \to M$ be a chart at P. Pick $t_0 > 0$ small enough that $h(L \times t_0) \subseteq B(P,M;\underline{\tau})$, and let f be the composite

$$(L \times t_0) \xrightarrow[h]{} B^{o}(P,M;\underline{\tau}) \xrightarrow[e_P^{-1}]{} (T_PM)^{o} \xrightarrow[(T_Ph)^{-1}]{} (c^{\infty}L)^{o},$$

where T_Ph is as in Theorem 2.26.2.

LEMMA 6.6. If t_0 is sufficiently small, then f and $id_{L\times t_0}$ are homotopic in $(c^{\infty}L)^{0}$.

Proof. Let $r_1 \in (0,1)$ be as in Lemma 3.8 and (3.8.8); then by (3.8.7), whenever $(Z,t),(Z',t') \in (c^{\infty}L)^{o}$ are distant less than tr_1, there is a unique geodesic $b_{(Z,t),(Z',t')}$ from (Z,t) to (Z',t'); moreover $im(b_{(Z,t),(Z',t')}) \subseteq (c^{\infty}L)^{o}$. Let $(Z,t_0) \in L \times t_0$ and let γ be the geodesic from P to $X = h(Z,t_0)$ parametrized by arc length with domain $[0,\tau_0]$, so $\tau_0 = d^M(P,h(Z,t_0))$. Set $g = h^{-1}\circ\gamma$, and for $\tau > 0$, write $g(\tau) = (Z(\tau),t(\tau))$. By Lemma 2.18.1, $t(\tau) = \tau(1 + O(\tau^{2k}))$. By Proposition 2.14 and (2.11) for γ,

$$\lim_{\tau\to 0^+} (Z(\tau),t(\tau)/\tau) = (Z^*,1) \text{ exists,}$$

thus defining a point $Z^* \in L$. (See Figure 10.) Now

$$\begin{aligned} f(Z,t_0) &= (T_Ph)^{-1}(\hat{\gamma}_X) \\ &= (T_Ph)^{-1}(\tau_0\hat{\gamma}), && \text{by (3.4.1)} \\ &= \tau_0 \lim_{\tau\to 0^+} (Z(\tau),t(\tau)/\tau), && \text{by the proof of Theorem 2.26} \\ &= (Z^*,\tau_0). \end{aligned}$$

By Lemma 2.21,

$$d^L(Z,Z^*) = O(\tau_0^{\,k}),$$

and by Lemma 2.18.2,

$$\tau_0 = t_0(1 + O(t_0^{2k})).$$

Combining these estimates with (1.16.5) gives

$$\begin{aligned} d^{c^\infty L}((Z,t_0),f(Z,t_0)) &\leq \{t_0^2 + \tau_0^2 - 2t_0\tau_0 \cos(d^L(Z,Z^*))\}^{1/2} \\ &= t_0 O(t_0^k) \\ &< t_0 r_1, \end{aligned}$$

once t_0 is small enough, independently of Z. For such t_0, $b_{(Z,t_0),f(Z,t_0)}$ exists and, since it is unique, varies continuously in (Z,t_0). So

$$B(Z,s) = b_{(Z,t_0),f(Z,t_0)}(s)$$

is a homotopy from $\mathrm{id}_{L\times t_0}$ to f, as required. q.e.d.

COROLLARY 6.6.1. Let $p\colon (c^\infty L)^o \to L$ be the natural projection; then $L \times t_0 \xrightarrow{f} (c^\infty L)^o \xrightarrow{p} L$ is surjective.

Proof. Since L is a closed, compact manifold, $p \circ \mathrm{id}_{L\times t_0}$ has non-zero degree (mod 2). Hence so does $p \circ f$, and the corollary follows. q.e.d.

LEMMA 6.7. $e_P^{-1}(B(P,M;\underline{\tau}))$ is a neighbourhood of $\hat{P}$ in $T_P M$.

Proof. I use the notation of the proof of Lemma 6.6. Define $F\colon cL \to c^\infty L$ by

$$F(Z,t) = \gamma_{h(Z,t_0)}(tt_0).$$

Then

$$F(Z,1) = f(Z,t_0) = (Z^*,\tau_0),$$

and so

$$F(Z,t) = (Z^*,t\tau_0).$$

Hence im(F) is the cone from c on im(f) in $c^\infty L$. By Corollary 6.6.1, im(F)

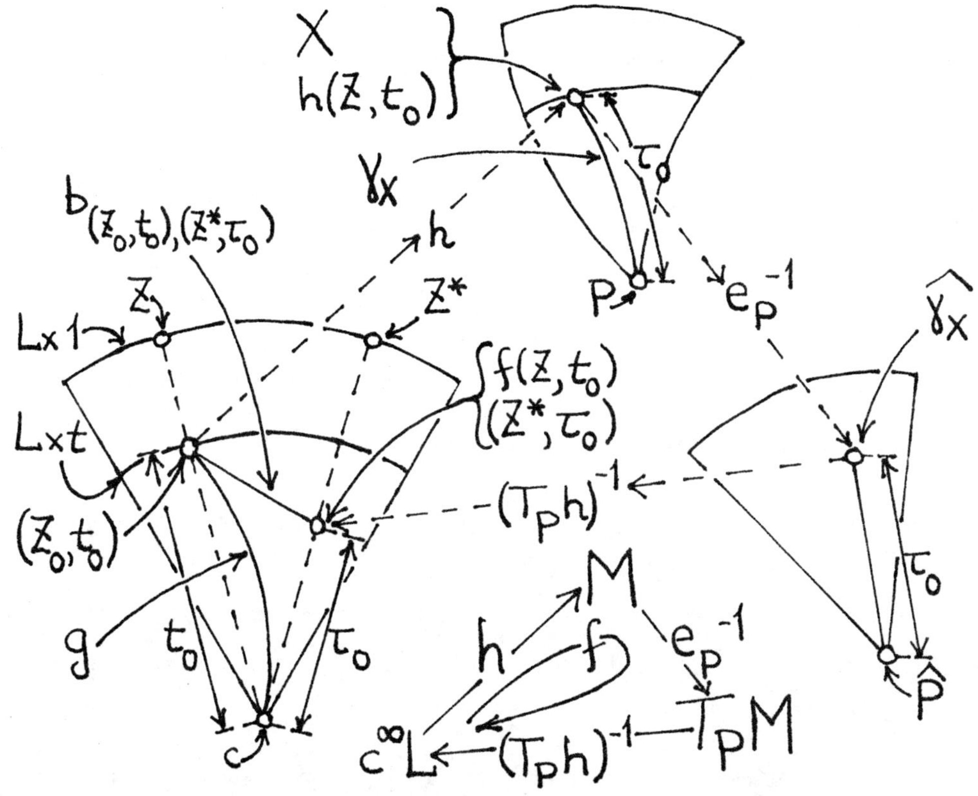

Figure 10

is a neighbourhood of c in $c^{\infty}L$. Now F factors through $B(P,M;\underline{\tau})$. Applying the homeomorphism $T_P h\colon\ c^{\infty}L \to T_P M$ completes the proof of the lemma. q.e.d.

PROPOSITION 6.8. $e_P^{-1}(B(P,M;\underline{\tau})) = B(\hat{P},T_P M;\underline{\tau})$.

Proof. Let $X \in B(P,M;\underline{\tau})$; then $e_P^{-1}(X) = \hat{\gamma}_X$, and

$$
\begin{aligned}
d^T(\hat{P},\hat{\gamma}_X) &= |\hat{\gamma}_X|, && \text{by } (2.25.2)\\
&= d^M(P,X), && \text{by } (3.4.1)\\
&< \underline{\tau}, && \text{by } (6.1.2).
\end{aligned}
$$

Thus

$$e_P^{-1}(B(P,M;\underline{\tau})) \subseteq B(\hat{P},T_P M;\underline{\tau}).$$

Now let $\hat{Y} \in B^{o}(\hat{P},T_{P}M;\underline{\tau})$. By Lemma 6.7 there exist $\lambda^* \in (0,1]$ and $X^* \in M^{o}$ such that $\hat{\gamma}_{X^*} = \lambda^* Y$. Since $|\gamma_{X^*}| < \lambda^*\underline{\tau}$, γ_{X^*} can be extended to a geodesic with domain $[0,1/\lambda^*]$, by Lemma 3.17. Set $X = \gamma_{X^*}(1/\lambda^*)$. Then by (3.4.2),

$$\hat{\gamma}_X = (\lambda^*)^{-1}\hat{\gamma}_{X^*} = \hat{Y}.$$

Thus

$$e_P^{-1}(B(P,M;\underline{\tau})) \supseteq B(\hat{P},T_PM;\underline{\tau}). \qquad \text{q.e.d.}$$

THEOREM 6.9. Let $\underline{\tau}$ be as in (5.1.1). Then there is a homeomorphism

$$e_P\colon\ B(\hat{P},T_PM;\underline{\tau}) \to B(P,M;\underline{\tau}),$$

defined by

$$\hat{\gamma}_{e_P\hat{Y}} = \hat{Y},$$

such that (1.1) -- (1.5) hold. Moreover e_P and e_P^{-1} both satisfy uniform Lipschitz conditions.

<u>Proof</u>. (1.1) follows from Corollary 2.17.4, (1.2) from Theorem 3.3, (1.4) from Proposition 6.8 and (1.5), with the additional Lipschitz conditions, from Corollary 6.3.5 and Proposition 6.4.2. q.e.d.

PROPOSITION 6.10. Let α and β be paths from P in M which are C^1 at P. Then

$$\lim_{\lambda\to 0^+} d^M(\alpha\lambda,\beta\lambda)/\lambda \text{ exists and equals } d^T(\alpha,\beta).$$

<u>Proof</u>. I first consider the special case:

(6.10.1) $|\hat{\alpha}|,\ |\hat{\beta}| < \underline{\tau}$.

Set $X = e_P(\hat{\alpha})$ and $Y = e_P(\hat{\beta})$; then γ_X and γ_Y are representatives of $\hat{\alpha}$ and $\hat{\beta}$ respectively. By Lemmas 6.3.1 and 6.3.2,

(6.10.2) $\displaystyle\lim_{\lambda\to 0^+} d^M(\gamma_X\lambda,\gamma_Y\lambda)/\lambda$ exists and equals $d^T(\hat{\alpha},\hat{\beta})$.

Since α and γ_X are equivalent,

$$d^M(\alpha\lambda,\gamma_X\lambda) = o(\lambda;\ \text{fixed}\ \alpha),$$

by (2.25). Similarly

$$d^M(\beta\lambda,\gamma_Y\lambda) = o(\lambda;\ \text{fixed}\ \beta).$$

Hence

$$\lim_{\lambda\to 0^+} (d^M(\alpha\lambda,\beta\lambda)/\lambda - d^M(\gamma_X\lambda,\gamma_Y\lambda)/\lambda)$$

exists and equals 0. The proposition now follows from (6.10.2), under the assumption (6.10.1).

In the general case, choose $r > 0$ so small that

$$r|\hat{\alpha}|,\ r|\hat{\beta}| < \underline{\tau}.$$

Define ${}_r\alpha$ by ${}_r\alpha(\mu) = \alpha(r\mu)$ and define ${}_r\beta$ similarly. Then ${}_r\alpha$ and ${}_r\beta$ are paths from P which are C^1 at P and which represent $r\hat{\alpha}$ and $r\hat{\beta}$ respectively. Now

$$\begin{aligned} d^T(\hat{\alpha},\hat{\beta}) &= r^{-1}d^T(r\hat{\alpha},r\hat{\beta}), && \text{by Theorem 2.26.1} \\ &= r^{-1} \lim_{\mu\to 0^+} d^M({}_r\alpha\mu,{}_r\beta\mu)/\mu, && \text{by the previous argument} \\ &= \lim_{\lambda\to 0^+} d^M(\alpha\lambda,\beta\lambda) \quad , \end{aligned}$$

setting $\lambda = r\mu$. In particular the last limit exists. q.e.d.

The Structure of T_PM

(6.11) Definitions. For the rest of this chapter $B^o(\hat{P},T_PM;\underline{\tau})$ will have the C^∞ structure induced by e_P^{-1} from M^o. Given $\hat{X} \in B^o(\hat{P},T_PM;\underline{\tau})$, set

(6.11.1) $X = e_P(\hat{X})$, so $\hat{X} = \hat{\gamma}_X$ and $|\hat{X}| = \tau(X)$,

by (2.25.3) and (3.4.1). Let $||_P$ be the norm on T_{γ_X} (see (4.14.1)) defined in Proposition 4.21. The induced norm on $T_{\hat{X}}(T_P M)$ is

$$(6.11.2) \quad |v| = |\underline{J}\{De_P(\hat{X})\cdot v\}|_P = \lim_{\lambda \to 0^+} |\underline{J}\{De_P(\hat{X})\cdot v\}(\lambda)|/\lambda,$$

for any $v \in T_{\hat{X}}(T_P M)$. (See Figure 11.) The corresponding inner product on $T_{\hat{X}}(T_P M)$ is

$$(6.11.3) \quad \langle v,w\rangle = (|v+w|^2 - |v|^2 - |w|^2)/2.$$

Let $\underline{\partial}_t$ be the vector field on $B^o(\hat{P}, T_P M; \underline{\tau})$ defined by

$$(6.11.4) \quad \underline{\partial}_t(\hat{X}) = De_P{}^{-1}(\hat{X})\cdot \underline{\partial}_\tau(X) = |\gamma_X{}'|^{-1} \frac{d}{dt}(e_P{}^{-1} \circ \gamma_X(t))|_{t=1}, \text{ by (3.4.3) and (3.4.4)} = |\hat{X}|^{-1} \frac{d}{dt}(t\hat{X})|_{t=1},$$

in view of (6.11.1). By (4.22) and (6.11.2),

$$|\underline{\partial}_t(\hat{X})| = 1;$$

thus $\underline{\partial}_t$ may be called the <u>unit radial vector field</u> on $B^o(\hat{P}, T_P M; \underline{\tau})$. More generally:

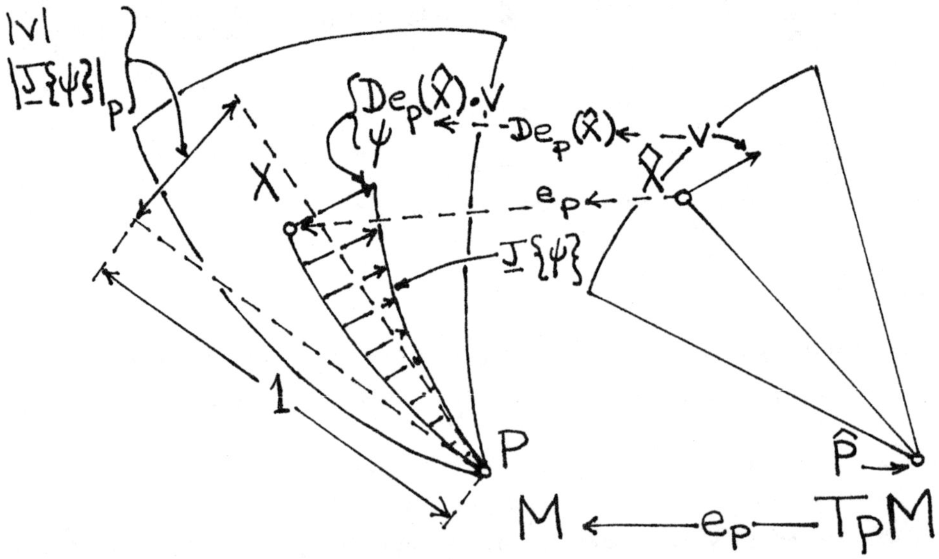

Figure 11

(6.11.5) if $v \in T_{\hat{X}}(T_P M)$ is a scalar multiple of $\underline{\partial}_t(\hat{X})$, then

$$|v| = |De_P(\hat{X})\cdot v|.$$

LEMMA 6.12. In the notation of (6.11),

$$\langle v, \underline{\partial}_t(\hat{X})\rangle = \langle De_P(\hat{X})\cdot v, \underline{\partial}_\tau(X)\rangle,$$

for any $v \in T_{\hat{X}}(T_P M)$.

Proof. Set $\psi = De_P(\hat{X})\cdot v$. First assume that v is a scalar multiple of $\underline{\partial}_t(\hat{X})$, so that ψ is a multiple of $\underline{\partial}_\tau(X)$ by (6.11.4). Then

$$\begin{aligned} 2\langle v, \underline{\partial}_t(\hat{X})\rangle &= |v+\underline{\partial}_t(\hat{X})|^2 - |v|^2 - |\underline{\partial}_t(\hat{X})|^2, \quad \text{by (6.11.3)} \\ &= |\psi+\underline{\partial}_\tau(X)|^2 - |\psi|^2 - |\underline{\partial}_\tau(X)|^2, \quad \text{by (6.11.5)} \end{aligned}$$

which proves the lemma in this special case.

In the general case, write

$$\psi = \psi^{\parallel} + \psi^{\perp},$$

where, in the notation of (1.13.5),

$$\psi^{\parallel} = \underline{\partial}_\tau(X)^{\parallel}(\psi), \quad \psi^{\perp} = \underline{\partial}_\tau(X)^{\perp}(\psi).$$

Let $\underline{J}$, ${}^{\parallel}\underline{J}$ and ${}^{\perp}\underline{J}$ be the Jacobi fields along γ_X from P with final conditions ψ, $\psi^{\parallel}$ and $\psi^{\perp}$ respectively, as in (4.11) and (4.14.2). Then by (4.11.7) and (4.11.9),

$$\langle {}^{\perp}\underline{J}(\lambda), \gamma_X{}'(\lambda)\rangle = 0,$$

$${}^{\parallel}\underline{J}(\lambda) \text{ is a scalar multiple of } \gamma_X{}'(\lambda), \text{ and}$$

$$\underline{J}(\lambda) = {}^{\parallel}\underline{J}(\lambda) + {}^{\perp}\underline{J}(\lambda),$$

for all $\lambda \in (0,1]$. Thus

$$\langle {}^{\parallel}\underline{J}(\lambda), {}^{\perp}\underline{J}(\lambda)\rangle \equiv 0,$$

and

(6.12.1) $|\underline{J}(\lambda)|^2 = |^{\parallel}\underline{J}(\lambda)|^2 + |^{\perp}\underline{J}(\lambda)|^2.$

Now set

$$v^{\parallel} = De_P{}^{-1}(X)\cdot\psi^{\parallel}\ ,\ v^{\perp} = De_P{}^{-1}(X)\cdot\psi^{\perp}.$$

Then $v = v^{\parallel} + v^{\perp}$, and $v^{\parallel}$ is a scalar multiple of $\underline{\partial}_t(\hat{X})$, by (6.11.4). It follows from (6.12.1) and (6.11.2) that

$$|v|^2 = |v^{\parallel}|^2 + |v^{\perp}|^2;$$

and therefore that

$$\langle v^{\parallel}, v^{\perp}\rangle = 0.$$

Hence

$$\langle v^{\perp}, \underline{\partial}_t(\hat{X})\rangle = 0.$$

So

$$\begin{aligned}\langle v, \underline{\partial}_t(\hat{X})\rangle &= \langle v^{\parallel}, \underline{\partial}_t(\hat{X})\rangle \\ &= \langle \psi^{\parallel}, \underline{\partial}_\tau(X)\rangle, \text{ as proved above} \\ &= \langle \psi, \underline{\partial}_\tau(X)\rangle.\end{aligned}$$

q.e.d.

THEOREM 6.13. e_P: $B(\hat{P}, T_PM; \underline{\tau}) \to M$ satisfies all the axioms of a chart except (1.6.2).

Proof. (1.6.1) is trivial. In (1.6.3), I is trivial, II holds by (6.11), III follows from (6.11.4) and IV from Lemma 6.12. Hypothesis V was proved as Lemma 5.18.2 (in view of (6.11.4)), VI is trivial, and VII was proved as Lemma 4.1 (again using (6.11.4)). q.e.d.

LEMMA 6.14. Let the notation be that of (6.11). Let $r \in (0,1]$ be given. Then:

(1) There exists a linear isomorphism

$$\partial r(\hat{X}):\ T_{\hat{X}}(T_PM) \to T_{r\hat{X}}(T_PM)$$

such that for every $v \in T_{\hat{X}}(T_PM)$, the directional derivative $\frac{\partial r}{\partial v}(\hat{X})$ of the dilation $r\colon T_PM \to T_PM$ in the direction v at $\hat{X}$ exists and satisfies

$$\frac{\partial r}{\partial v}(\hat{X}) = \partial r(\hat{X})\cdot v;$$

(2) If also $s \in (0,1]$, then

$$\partial s(r\hat{X})\cdot\partial r(\hat{X}) = \partial(rs)(\hat{X});$$

(3) $|\partial r(\hat{X})\cdot v| = r|v|$.

Proof of (1). Let $\hat{\beta}$ be a C^∞ curve in $B^o(\hat{P},T_PM;\underline{\tau})$ such that

(6.14.4) $\hat{\beta}(0) = \hat{X}$ and $\hat{\beta}'(0) = v$.

Set $\beta = e_P \circ \hat{\beta}$; then β is a C^∞ curve in $B^o(P,M;\underline{\tau})$ from X. By Theorem 5.13 the curve $\beta^r(\sigma) = \gamma_{\beta\sigma}(r)$ is differentiable, and $\beta^{r\prime}(0) = \underline{J}(r)$, where $\underline{J}$ is the Jacobi field along γ_X from P with final condition $\beta'(0)$. Set $\hat{\beta}^r = e_P^{-1}\circ\beta^r$; then $\hat{\beta}^r$ is differentiable and by (3.4.2), $\hat{\beta}^r(\sigma) = r\hat{\beta}(\sigma)$. Now

(6.14.5) $\hat{\beta}^{r\prime}(0) = De_P^{-1}(\gamma_X r)\cdot\underline{J}(r)$

is independent of the choice of $\hat{\beta}$ satisfying (6.14.4). Therefore

(6.14.6) $\frac{\partial r}{\partial v}(\hat{X})$ exists and equals $\hat{\beta}^{r\prime}(0)$.

In the notation of (4.14),

(6.14.7) $\underline{J} = \underline{J}\{\beta'(0)\}$
$\qquad = \underline{J}\{De_P(\hat{X})\cdot v\}$.

By (6.14.5) and (6.14.6),

$$\frac{\partial r}{\partial v}(\hat{X}) = De_P^{-1}(\gamma_X r)\cdot(\underline{J}\{De_P^{-1}(\hat{X})\cdot v\}(r)).$$

The right-hand side here defines a linear isomorphism (in v)

$$\partial r(\hat{X})\colon\ T_{\hat{X}}(T_PM) \to T_{r\hat{X}}(T_PM),$$

because of Lemma 4.20. This proves (1).

Proof of (2). Let $\underline{J}\{\beta^{r}{}'(0)\}$ be the Jacobi field along $\gamma_{\beta^r 0}$ from P with final condition $\beta^{r}{}'(0)$, as in (4.14). It is sufficient to prove

(6.14.8) $\underline{J}\{\beta^{r}{}'(0)\}(s) = \underline{J}(rs)$,

where $\underline{J}$ is as in (6.14.7). Since

$$\gamma_{\beta^r 0}(\mu) = \gamma_X(r\mu),$$

for all $\mu \in [0,1]$, the vector field $\underline{J}^*$ defined by

$$\underline{J}^*(\mu) = \underline{J}(r\mu)$$

is a Jacobi field along $\gamma_{\beta^r 0}$ from P, with final condition $\beta^{r}{}'(0)$. It follows from Theorem 4.13.3 that

$$\underline{J}^* = \underline{J}\{\beta^{r}{}'(0)\},$$

which proves (6.14.8), and with it, assertion (2).

Proof of (3).

$$\begin{aligned} |\partial r(\hat{X})\cdot v| &= |\beta^{r}{}'(0)|, && \text{by (6.14.6)} \\ &= \lim_{\mu\to 0^+} |\underline{J}\{\beta^{r}{}'(0)\}(\mu)|/\mu, && \text{by (6.14.2)} \\ &= \lim_{\mu\to 0^+} |\underline{J}\{\beta'(0)\}(r\mu)|/\mu, && \text{by (6.14.8) and (6.14.7)} \\ &= r \lim_{\lambda\to 0^+} |\underline{J}\{\beta'(0)\}(\lambda)|/\lambda, && \text{setting } \lambda = r\mu \\ &= r|v|, && \text{by (6.11.2) again.} \end{aligned}$$

q.e.d.

THEOREM 6.15. The metric d^T on T_PM is intrinsic in the sense of Aleksandroff (see (2.5.5)).

Proof. It is required to prove that for all $\hat{X},\hat{Y} \in T_PM$,

$$d^T(\hat{X},\hat{Y}) = \inf\{L^*(\hat{\delta})\}$$

$$= d^*(\hat{X},\hat{Y}),$$

the infimum being taken over all continuous paths $\hat{\delta}$ from $\hat{X}$ to $\hat{Y}$ in T_PM. Thus

$$L^*(\hat{\delta}) = \sup_{\Pi} \sum_{i=1}^{m} d^T(\hat{\delta}\sigma_{i-1},\hat{\delta}\sigma_i),$$

where Π ranges over all partitions of the domain [0,1] of $\hat{\delta}$:

$$\Pi:\quad 0 = \sigma_0 < \sigma_1 < \ldots < \sigma_m = 1.$$

It follows from Theorem 2.26.1 that for any $r > 0$, $d^T(r\hat{X},r\hat{Y}) = rd^T(\hat{X},\hat{Y})$, and $L^*(r\hat{\delta}) = rL^*(\hat{\delta})$. Hence it is enough to prove the theorem under the additional assumption

(6.15.1) $\quad \hat{X},\hat{Y} \in B(\hat{P},T_PM;\underline{\tau})$.

Also $d^T(\hat{X},\hat{Y}) \le d^*(\hat{X},\hat{Y})$ by (2.5.6); hence it is necessary only to prove

(6.15.2) $\quad d^T(\hat{X},\hat{Y}) \ge d^*(\hat{X},\hat{Y})$.

I shall first prove (6.15.2) in some elementary cases.

Case I. $\hat{Y} = \hat{P}$. Then $d^T(\hat{X},\hat{P}) = |\hat{X}|$. Define $\hat{\delta}(\sigma) = \sigma\hat{X}$. Write $X = e_P(\hat{X})$; then by (6.11.1) and (3.4.2),

$$\begin{aligned}\hat{\delta}(\sigma) &= \sigma\hat{\gamma}_X \\ &= \hat{\gamma}_{\gamma_X\sigma}.\end{aligned}$$

Also

$$\gamma_{\gamma_X\sigma}(\lambda) = \gamma_X(\lambda\sigma),$$

again by (3.4.2). It follows that for any Π,

$$\begin{aligned}\sum_{\Pi}(\hat{\delta}) &= \sum_{i=1}^{m} d^T(\hat{\delta}\sigma_{i-1},\hat{\delta}\sigma_i) \\ &= \sum_{i=1}^{m} \lim_{\lambda\to 0^+} d^M(\gamma_X(\lambda\sigma_{i-1}),\gamma_X(\lambda\sigma_i))/\lambda,\end{aligned}$$

by Proposition 6.10

$$= \sum_{i=1}^{m} d^M(\gamma_{X}\sigma_{i-1}, \gamma_{X}\sigma_i)$$

$$= d^M(P,X)$$

$$= \tau(X)$$

$$= |\hat{X}|, \qquad \text{by (6.11.1)}$$

$$= d^T(\hat{X},\hat{P}).$$

Thus

$$d^*(\hat{X},\hat{Y}) \leq L^*(\hat{\delta}) = d^T(\hat{X},\hat{P}),$$

which proves (6.15.2) in Case I.

<u>Case II</u>. $d^T(\hat{X},\hat{Y}) = d^T(\hat{X},\hat{P}) + d^T(\hat{P},\hat{Y})$. This case follows directly from Case I.

<u>Case III</u>. The remaining possibility is

$$\text{(6.15.3)} \quad d^T(\hat{X},\hat{Y}) < d^T(\hat{X},\hat{P}) + d^T(\hat{P},\hat{Y}) = |\hat{X}| + |\hat{Y}|.$$

The proof of (6.15.2) in this case will be given in (6.17), after the next lemma.

LEMMA 6.16. Let β be a differentiable path in $B^o(P,M;\underline{\tau})$. Set $\hat{\beta} = e_P^{-1} \circ \beta$, and for each $\mu \in (0,1]$ define β^μ by $\beta^\mu(\sigma) = \gamma_{\beta\sigma}(\mu)$. Then

$$\lim_{\mu \to 0^+} L^*(\beta^\mu)/\mu \text{ exists and equals } L^*(\hat{\beta}).$$

<u>Proof</u>. Fix a partition Π of $[0,1]$ and $\lambda \in (0,1]$. By Lemma 6.2, for $\mu \in (0,\lambda]$,

$$d^M(\gamma_{\beta\sigma_{i-1}}\mu, \gamma_{\beta\sigma_i}\mu)/\mu$$

$$= (1 + O((\lambda\underline{\tau})^{1/q}))d^M(\gamma_{\beta\sigma_{i-1}}\lambda, \gamma_{\beta\sigma_i}\lambda)/\lambda.$$

Let $\mu \to 0^+$; then by Lemmas 6.3.1 and 6.3.2,

$$d^T(\hat{\beta}\sigma_{i-1},\hat{\beta}\sigma_i) = (1 + O((\lambda\underline{\tau})^{1/q}))d^M(\gamma_{\beta\sigma_{i-1}}\lambda,\gamma_{\lambda\sigma_i}\lambda)/\lambda.$$

Summing over $i = 1, \ldots, m$ gives

$$\sum_{\Pi}(\hat{\beta}) = (1 + O((\lambda\underline{\tau})^{1/q}))(\sum_{\Pi}(\beta^\lambda))/\lambda.$$

Since the term $1 + O(\lambda\underline{\tau})^{1/q}$ is independent of Π, I may take the supremum over Π. This gives

$$L^*(\hat{\beta}) = (1 + O((\lambda\underline{\tau})^{1/q}))L^*(\beta^\lambda)/\lambda.$$

The lemma now follows by letting $\lambda \to 0^+$. q.e.d.

(6.17) Proof of Theorem 6.15, concluded. It remains to prove (6.15.2) under the assumptions (6.15.1) and (6.15.3). Set $X = e_P(\hat{X})$, $Y = e_P(\hat{Y})$. It follows from (6.15.3) and the definition (2.25.1) of d^T that for all sufficiently small $\lambda \in (0,1]$,

$$d^M(\gamma_X\lambda,\gamma_Y\lambda) < d^M(\gamma_X\lambda,P) + d^M(P,\gamma_Y\lambda).$$

For any such λ, let β^λ be a geodesic from $\gamma_X\lambda$ to $\gamma_Y\lambda$; then, in view of (6.15.1) and (6.1.1), $\mathrm{im}(\beta^\lambda) \subseteq B^O(P,M;\lambda\underline{\tau})$. By Proposition 5.17 there exists a differentiable path $\beta[\lambda]$ from X to Y in $B^O(P,M;\underline{\tau})$ such that

$$\gamma_{\beta[\lambda]}(\sigma) = \beta^\lambda(\sigma),$$

for all σ. (See Figure 12.) Set

$$\hat{\beta[\lambda]} = e_P^{-1}\circ\beta[\lambda].$$

Now by (6.2.1), for $\mu \in (0,\lambda]$,

$$L^*(\beta[\lambda]^\mu) = (\mu/\lambda)(1 + O((\lambda\underline{\tau})^{1/q}))L(\beta^\lambda).$$

Hence

$$L^*(\beta[\lambda]^\mu)/\mu = (1 + O((\lambda\underline{\tau})^{1/q}))d^M(\gamma_X\lambda,\gamma_Y\lambda)/\lambda.$$

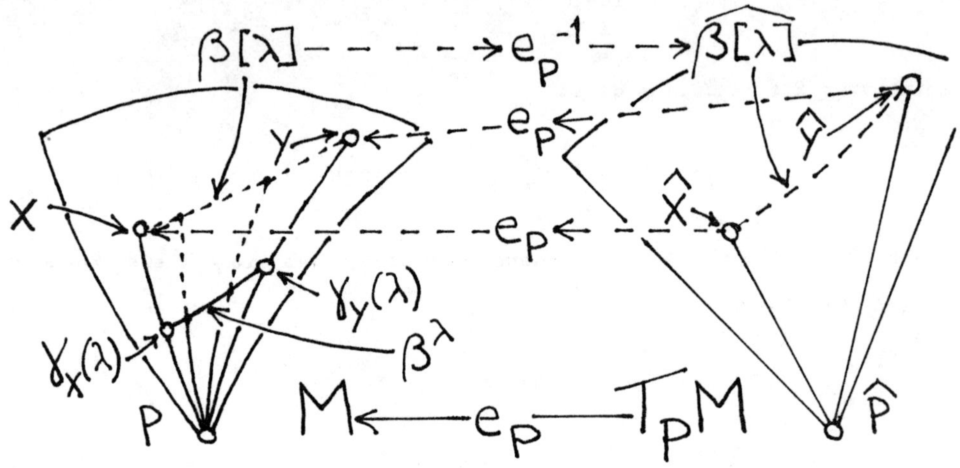

Figure 12

By Lemma 6.16, taking the limit as $\mu \to 0^+$ gives

$$L^*(\hat{\beta[\lambda]}) = (1 + O((\lambda\underline{\tau})^{1/q}))d^M(\gamma_X\lambda,\gamma_Y\lambda)/\lambda.$$

As $\lambda \to 0^+$, the right-hand side converges to $d^T(\hat{X},\hat{Y})$, by (2.25.1). Therefore

$$\lim_{\lambda\to 0^+} L^*(\hat{\beta[\lambda]}) = d^T(\hat{X},\hat{Y}).$$

In view of (2.5.4), this proves (6.15.2), and with it, Theorem 6.15.

q.e.d.

CHAPTER 7: AFFINE COMPLEX ANALYTIC HYPERSURFACES

Introduction

In this chapter I prove that for affine complex analytic hypersurfaces $M^n \subseteq \mathbb{C}^{n+1}$ certain types of isolated singularity in the sense of Milnor [9] are conical in the sense of (1.6), so that M has an exponential map at such points P by Theorem 6.9. The results look rather different according as n = 1 or n > 1. In either case a chart h: $c\underline{t}L \to M$ at P is defined by modifying as little as possible the natural structure of $\mathbb{C}^{n+1}$ as cone from P. Namely, h is defined (Lemma 7.5) so that

$$L \times \underline{t} = \{X \in M \text{ such that } |X| = \underline{t}\};$$
$$|h(Z,t)| = t;$$

and, setting $X = h(Z,t)$,

$\frac{\partial h}{\partial t}(X)$ is a real-scalar multiple of the orthogonal projection into $T_X M$ of the radial vector $\overrightarrow{PX}$.

In verifying the axioms of a chart it is V and VII of (1.6.3) that are most difficult; and one measure of the inherent difficulty involved is the care with which k must be chosen in these two axioms. When n = 1, any singular point is an isolated conical singularity (Theorem 7.11, which proves Theorem 1.9). It is easy to reduce the theorem to the case that M has equation

$$f(x,y) = ax^q + \sum_{i=q+1}^{\infty} a_i x^i - y^p = 0, \text{ where } q > p \text{ and } a \neq 0.$$

Then one must choose

$$k < (1/2)\min(1,q/p-1).$$

Thus there exist curves for which k has to be arbitrarily small. When n > 1 a complication arises that cannot occur when n = 1. Say M has equation $f = 0$ and let g be the initial homogeneous polynomial of f, so that $T = \{g = 0\}$ is the Whitney tangent cone of M at P. Then (for $n > 1$) it is possible for grad g (see (7.1.3) or Milnor [9]) to be $\bar{0}$ at points of M arbitrarily close to P. IN such cases M can fail to have an exponential map at P((Example 7.23). To avoid this possibility I make the strong assumption that P is an unbranched singularity of M; that is, that on all of $\mathbb{C}^{n+1}$, grad g is $\vec{0}$ only at P. Theorem 7.6 proves that if P is an unbranched singularity of M, then all the hypotheses of (1.6.3) hold except VI, the upper bound on sectional curvature (Examples in which VI definitely fails are given in 7.24.) Theorem 7.6 implies that the results of Chapter 2 hold; in particular (Corollary 7.6.1) that T_PM is homeomorphic to M near P and that every geodesic from P is C^1 at P. Theorem 1.8 also follows immediately from Theorem 7.6. The strength of the hypothesis that P be unbranched in Theorem 7.6 may be assessed from the fact that it is enough to choose $k = 1/2$ in V and VII of (1.6.3).

It is of interest to identify T_PM and its metric d^T. This can be done when $n = 1$. Again consider the special case that M has the equation above, so that T is the x-axis. Let $\pi\colon M \to T$ be orthogonal projection. Its restriction $\pi^o\colon M^o \to T^o$ is a p-fold covering map near P. There is an induced map $\hat{\pi}\colon T_PM \to T$. Theorem 7.18 shows that $\hat{\pi}$ is a p-fold covering map branched at P, which is a local isometry on $(T_PM)^o$. This theorem leads to a complete description of T_PM for any curve M (Corollary 7.18.4). It also interprets geometrically the fact that the algebraic tangent cone of M at P is T taken with multiplicity p.

When $n > 1$, Theorem 7.10 shows that if P is unbranched, then T_PM is homeomorphic to T, by using an equisingularity theorem of Teissier [15] to prove that M and T are homeomorphic near P and applying Corollary 7.6.1. The rest is conjectural. In (7.2.4.1) I conjecture that $\exp_P$ exists whenever P is an unbranched singularity of M. More generally, assume the weaker condition that the projective tangent cone of M at P, namely $T^o/\mathbb{C}^o$, is non-singular in $\mathbb{C}P^n$. Then there is again an "orthogonal projection" $\pi\colon M \to T$ near P defined by assigning to X that point $\pi(X) \in T$ which is closest to X. The restriction $\pi^o\colon M^o \to T^o$ is now a branched covering map in general. But when P is unbranched, π^o should be single-sheeted and therefore have no branching, so that π should be a homeomorphism. It is because of this line of thought that I have used the word "unbranched" to describe P in Theorem 7.6. Even assuming this reasoning about π to be correct, I do not see how to derive from it what is really needed: the corresponding results about the induced map $\hat{\pi}$: $T_P M \to T$. I conjecture that when P is unbranched, $\hat{\pi}$ is an isometry; this would describe $T_P M$ completely (see (7.10.5)).

Finally some examples and conjectures suggest what may lie ahead in trying to define an exponential map at a more general type of singular point of an affine complex analytic hypersurface. (n (7.22), P is not an isolated singularity and M cannot have an exponential map at P. In (7.23), as already mentioned, the same is true of M and P is isolated but not unbranched. For (7.23) to work it is essential that the "branch locus"

$$B = \{X \in M \text{ such that } (\underline{\text{grad}}\ g)(X) = \vec{0}\}$$

is not a cone. In (7.25), P is again an isolated singularity, but not unbranched, and now B is a cone. Here P fails to be a conical singularity, yet I believe that M nevertheless has an exponential map at P. Indeed (7.26) and (7.27) are conjectures that would unify the separate developments of this chapter for $n = 1$ and $n > 1$: they propose that if P is an isolated singular point and the branch locus is a cone, then M has an exponential map at P, and that the natural map $\hat{\pi}$: $T_P M \to T$ is a covering map and a local isometry on the complement of B. (7.28) raises a question that could lead to answers to some of the conjectures of this chapter. Let M be a complex analytic hypersurface with an isolated critical point P, and let h: $\mathbb{C}^{n+1} \to \mathbb{C}^{n+1}$ be an analytic isomorphism defined near P. Assume that $M^* = h(M)$ has an exponential map at $P^* = h(P)$. Is the same true for M at P? If so, Conjecture 7.24.1 would follow by choosing h so that $h(M) = T$, the Whitney tangent cone of M at P.

Hypersurfaces of Arbitrary Dimension

(7.1) Notation and definitions. When $\mathbb{C}^{n+1}$ is regarded as a manifold, its points will be denoted $X = (X_1,\ldots,X_{n+1})$ and its origin P. When $\mathbb{C}^{n+1}$ is thought of as a vector space, for example when identified with $T_X\mathbb{C}^{n+1}$, its vectors will be denoted v or w and its origin $\vec{0}$.

I shall write

$$X_i = \xi_i + \sqrt{-1}\eta_i$$

in real and imaginary parts. Let the vector fields $\underline{\partial}_i$, $\underline{\bar{\partial}}_i$ and $\underline{R}$ be defined on $\mathbb{C}^{n+1}$ by

(7.1.1) $$\underline{\partial}_i = \frac{\partial}{\partial X_i} = (1/2)\left(\frac{\partial}{\partial \xi_i} - \sqrt{-1}\,\frac{\partial}{\partial \eta_i}\right);$$

$$\underline{\bar{\partial}}_i = \frac{\partial}{\partial \bar{X}_i} = (1/2)\left(\frac{\partial}{\partial \xi_i} + \sqrt{-1}\,\frac{\partial}{\partial \eta_i}\right); \text{ and}$$

(7.1.2) $$\underline{R}(X) = X = \sum_1^{n+1} (X_i\underline{\partial}_i + \bar{X}_i\underline{\bar{\partial}}_i)(X).$$

(For the first two, see Wells [17, p. 35].) In these formulae and henceforth the superscript bar denotes complex conjugation.

Let f be a holomorphic function on some neighbourhood U of P in $\mathbb{C}^{n+1}$, and set $M = f^{-1}(0)$. Following Milnor [9], the vector field $\underline{\text{grad}}$ f is defined on U by

(7.1.3) $$\underline{\text{grad}}\ f = (\overline{\underline{\partial}_1 f}, \ldots, \overline{\underline{\partial}_{n+1} f}).$$

Then the derivative of f in the direction $v \in T_X\mathbb{C}^{n+1}$ is given by

(7.1.4) $$Df \cdot v = \underline{\langle} v, \underline{\text{grad}}\ f \underline{\rangle},$$

where $\underline{\langle}\ ,\ \underline{\rangle}$ is the standard Hermitian product on $\mathbb{C}^{n+1}$ (compare (1.13.2))

(7.1.5) $$\underline{\langle} v,w \underline{\rangle} = \sum_{i=1}^{n+1} v_i\bar{w}_i.$$

Let g be the initial polynomial of f; so g is homogeneous of degree deg g = p, say. Set h = f - g; then deg h > p. I assume henceforth that $P \in M$; equivalently, that $p \geq 1$. The <u>Whitney tangent cone of M at P</u> is

(7.1.6) $$T = g^{-1}(0).$$

The case p = 1 is not of interest in this work; for then M is an analytic submanifold of $\mathbb{C}^{n+1}$ near P, T is the tangent plane to M at P; and it is well-known that M has an exponential map at P which satisfies (1.1) -- (1.5). So for the rest of this section I shall assume:

(7.1.7) $\deg g = p \geq 2$.

This implies that

$$\underline{\text{grad}}\ g(P) = \underline{\text{grad}}\ f(P) = \vec{0}.$$

The <u>branch locus of M at P</u> is defined to be

(7.1.8) $B = \{X \in M \text{ such that } \underline{\text{grad}}\ g(X) = \vec{0}\}$.

P is called an <u>unbranched isolated singular point of M</u> if

(7.1.9) $(\underline{\text{grad}}\ g)^{-1}(\vec{0}) = \{P\}$.

It follows from Lemma 7.2.1 (below) that if (7.1.9) holds, then the domain U of f can be chosen so that

(7.1.10) $(\underline{\text{grad}}\ f)^{-1}(\vec{0}) = \{P\}$ on U;

so that P is an isolated singular point of M. From (7.1.10) follows that $M^o = M - \{P\}$ is a C^∞ submanifold of $\mathbb{C}^{n+1}$. It will be assumed henceforth that when (7.1.10) holds,

(7.1.11) M^o has the Riemannian metric induced from that of (7.1.5) and (1.13.3) on $\mathbb{C}^{n+1}$.

Again assuming (7.1.10) holds, define a C^∞ vector field $\underline{R}^{\perp}$ on U^o by

(7.1.12) $\underline{R}^{\perp}(X) = (\underline{\text{grad}}\ f)^{\perp}\underline{R}(X)$;

here I am using the notation of (1.13.6). Thus $\underline{R}^{\perp}\restriction M^o$ is tangent to M^o. Lemma 7.2.3 will show that $\underline{R}^{\perp}$ is nowhere zero on M^o.

LEMMA 7.2. Let the notation be as above, and assume that P is an unbranched isolated singular point of M. Then the domain U of f can be chosen small enough that:

(1) $|\underline{\text{grad}}\ f|^{-1}(X) = O(|X|^{-p+1})$ on U^0;

(2) $|\langle \underline{R}, \underline{\text{grad}}\ f \rangle|(X) = O(|X|^{p+1})$ on M^o;

(3) $|\underline{R}^{\perp}(X)| = |X|(1 + O(|X|^2))$ on M^o

Proof of (1). To start with, let U be any domain of f. By (7.1.9),

$$g = \inf\{|\underline{\text{grad}}\ g|(X) \text{ for } |X| = 1\}$$

is strictly positive. Now $\underline{\text{grad}}\ g$ is homogeneous of degree p - 1; hence for every $X \in U$,

$$|\underline{\text{grad}}\ g|(X) \geq |X|^{p-1}g.$$

On the other hand, since deg h $\geq$ p + 1, there exists $h > 0$ such that

$$|\underline{\text{grad}}\ h|(X) \leq h|X|^p.$$

Hence if $U \subseteq B(P,\mathbb{C}^{n+1}; g/(2h))$, then

$$\begin{aligned} |\underline{\text{grad}}\ f|(X) &\geq ||\underline{\text{grad}}\ g|(X) - |\underline{\text{grad}}\ h|(X)| \\ &\geq |X|^{p-1}(g - h|X|) \\ &\geq (g/2)|X|^{p-1}. \end{aligned}$$

Assertion (1) now follows.

Proof of (2). Write $h = \sum_{p+1}^{\infty} h_m$ as the decomposition of h into homogeneous polynomials, with deg $h_m = m$. From (7.1.2) and (7.1.4), and since f is holomorphic,

$$\begin{aligned} \langle \underline{R}, \underline{\text{grad}}\ f \rangle(X) &= \sum_1^{n+1} X_i \underline{\partial}_i f(X) \\ &= pg + \sum_{p+1}^{\infty} m\, h_m, \text{ by Euler's theorem} \\ &= pf + \sum_{p+1}^{\infty} (m-p) h_m \\ &= \sum_{p+1}^{\infty} (m-p) h_m \text{ on } M^o. \end{aligned}$$

Thus on M^o, $\langle \underline{R}, \underline{\text{grad}}\ f \rangle$ has order $\geq p+1$; so (2) follows.

Proof of (3). By (7.1.12) and by (1.13.6) with $v = \underline{\text{grad}}\ f(X)$,

$$(7.2.4)\quad |\underline{R} - \underline{R}^{\perp}|(X) = (|\underline{\text{grad}}\ f|^{-1}|\langle \underline{R}, \underline{\text{grad}}\ f \rangle|)(X)$$
$$= o(|X|^2),$$

by (1) and (2). Since $|\underline{R}(X)| = |X|$ by (7.1.2), assertion (3) now follows from the Pythagorean theorem. q.e.d.

(7.3) Notation. Assume henceforth that $U = B(P, \mathbb{C}^{n+1}; 2t)$, where $\underline{t} > 0$ is such that:

(7.3.1) $\underline{\text{grad}}\ f$ is nowhere zero on U^o; and $\underline{R}^{\perp}$ is nowhere zero on $M^o \cap U$.

Let $\$(\underline{t})$ be the sphere of radius $\underline{t}$ about P, and set

$$(7.3.2)\quad L = M \cap \$(\underline{t}).$$

Let $\underline{\rho}$ be the vector field on M^o defined by

$$(7.3.3)\quad \underline{\rho}(X) = (|\underline{R}|\,|\underline{R}^{\perp}|^{-2})(X)\,\underline{R}^{\perp}(X).$$

For each $Z \in L$, let α_Z be the integral curve of $\underline{\rho}$ satisfying

(7.3.4) $\alpha_Z(\underline{t}) = Z$, $\alpha_Z'(t) = \underline{\rho}(\alpha_Z t)$ (See Figure 13.)

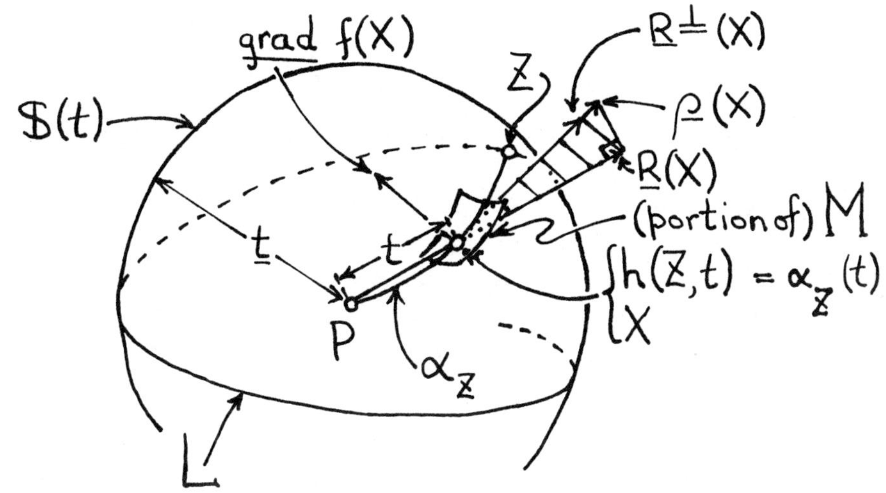

Figure 13

LEMMA 7.4. L is a C^∞ submanifold of $\$(\underline{t})$ of real codimension one in M.

Proof. The normal bundle of M^o in $\mathbb{C}^{n+1}$ has as fibre at $X \in M^o$ the $\mathbb{C}$-vector space generated by $\underline{\text{grad}}\ f(X)$. The normal bundle of $\$(\underline{t})$ in $\mathbb{C}^{n+1}$ has as fibre at $X \in \$(\underline{t})$ the $\mathbb{R}$-vector space generated by $\underline{R}(X)$. By (7.3.1), $\underline{R}(X)$ and $\underline{\text{grad}}\ f(X)$ are linearly independent over $\mathbb{C}$ for all $X \in M^o$. Therefore $\$(\underline{t})$ is transverse to M^o, and so L is a submanifold of $\$(\underline{t})$ and has real codimension 1 in M. Since M^o and $\$(\underline{t})$ are themselves C^∞ submanifolds of $\mathbb{C}^{n+1}$, it follows that L is a C^∞ submanifold of $\$(\underline{t})$. q.e.d.

LEMMA 7.5. (1) For every $Z \in L$, α_Z can be defined by (7.3.4) on the domain $(0,\underline{t}]$.

(2) Define $h\colon\ L \times (0,\underline{t}] \to M^o$ by $h(Z,t) = \alpha_Z(t)$; then h is a C^∞ diffeomorphism;

(3) For each $Z \in L$, $\lim_{t\to 0^+} h(Z,t) = P$, and the convergence is uniform in Z.

Proof. Let $Z \in L$ be arbitrary; then

$$\begin{aligned}
\frac{d}{dt}|\alpha_Z t| &= \langle \alpha_Z{}'t, (|\underline{R}|^{-1}\underline{R})(\alpha_Z t)\rangle \\
&= \mathrm{Re}(\langle \underline{\rho}, |\underline{R}|^{-1}\underline{R}\rangle(\alpha_Z t)), \quad \text{by (1.13.3) and (7.3.4)} \\
&= (|\underline{R}^{\perp}|^{-2}\mathrm{Re}\langle \underline{R}^{\perp}, \underline{R}\rangle)(\alpha_Z t), \quad \text{by (7.3.3)} \\
&= (|\underline{R}^{\perp}|^{-2}\mathrm{Re}\langle \underline{R}^{\perp}, \underline{R}^{\perp}\rangle)(\alpha_Z t), \quad \text{by (7.1.12)} \\
&= 1.
\end{aligned}$$

In view of (7.3.4), on any domain of definition $(\varepsilon,\underline{t}]$ of α_Z with $\varepsilon \in (0,\underline{t})$,

(7.5.4) $|\alpha_Z t| = t.$

Now (1) follows from the theory of differential equations.

Proof of (2). It follows from (1) that h can be defined on the given domain. Now L is a C^∞ submanifold of M^o by Lemma 7.4, and $\underline{\rho}$ is a C^∞ vector field on M^o. It is known from the theory of differential equations that the solution $\alpha_Z(t)$ of (7.3.4) varies in C^∞ fashion in its parameters Z and t, which proves (2).

Proof of (3). This is an immediate consequence of (7.5.4). q.e.d.

COROLLARY 7.5.5. There is a continuous map h: $c\underline{t}L \to M$ defined by $h(c) = P$ and $h(Z,t) = \alpha_Z(t)$ for $Z \in L$, $t \in (0,\underline{t}]$.

THEOREM 7.6. Let P be an unbranched isolated singular point of a complex analytic hypersurface M. Let M^o have the Riemannian metric of (7.1.11). Then I -- V and VII of (1.6.3) hold at P, using as chart for M at P the map h: $c\underline{t}L \to M$ of Corollary 7.5.5.

COROLLARY 7.6.1. Let M and P be as in Theorem 7.6. Then T_PM is homeomorphic to M near P. Every geodesic γ from P is C^1 at P in the sense of (2.11), and thus determines a point $\hat{\gamma} \in T_PM$.

Proof. This follows from Theorems 7.6 and 2.26.2 and from Propositions 2.9 and 2.10. q.e.d.

CORLLARY 7.6.2. Let M and P be as in Theorem 7.6, and assume that (1.6.3, VI) holds (that is, that the sectional curvature of M^o is bounded above near P). Then M has an exponential map at P satisfying (1.1) -- (1.5)

Proof. This follows from Theorems 7.6 and 6.9. q.e.d.

Proof of Theorem 7.6. The proof incorporates Lemma 7.8 and is concluded in (7.9).

(1.6.1) follows from Lemma 7.2.1, as was explained under (7.1.10), and (1.6.2) was proved in Lemma 7.4. It remains to verify I -- V and VII of (1.6.3).

Proof of I. This is included in Corollary 7.5.5.

Proof of II. This was proved as Lemma 7.5.2.

Proof of III. Set $X = h(Z,t)$. Then

$$(7.6.3)\quad (Dh\cdot\underline{\partial}_t)(h(Z,t)) = \underline{\rho}(X),$$

by (7.3.4) and the definition of h in Lemma 7.5.2. So

(7.6.4) $|(Dh\cdot\underline{\partial}_t)(h(Z,t))| = |\underline{\rho}(X)|$

$= (|\underline{R}|/|\underline{R}^{\perp}|)(X),$ by (7.3.3)

$= |X|/(|X|(1 + O(|X|)),$ by (7.1.2) and Lemma 7.2.3

$= 1 + O(|X|)$

$= 1 + O(t),$ by (7.5.4).

Now III follows.

<u>Proof of IV</u>. Given $t \in (0,\underline{t}]$ and $u \in T_{(Z,t)}(L \times t)$, let b be a curve in $L \times t$ such that $b(0) = (Z,t)$ and $b'(0) = u$. Set $\beta = h\circ b$; then by (7.5.4), $\mathrm{im}(\beta) \subseteq \$(t)$, and

(7.6.5) $\beta'(0) = Dh\cdot u.$

Set $X = h(Z,t)$. Then since $\underline{R}(X)$ is normal to $\$(t)$ at X,

$$\langle \beta'0, \underline{R}(X)\rangle = 0.$$

Also $\beta'(0)$ is tangent to M; so

$$\langle \beta'0, \underline{\text{grad}}\ f(X)\rangle = 0.$$

In view of (7.1.12) and (7.3.3), it follows that

$$\langle \beta'0, \underline{\rho}(X)\rangle = 0.$$

Now IV follows from (7.6.3) and (7.6.5).

<u>Proof of V</u>. This is deferred until after Lemma 7.8.

<u>Proof of VII</u>. This is postponed until after Lemma 7.8.

(7.7) Let D^M and $D^{\mathbb{C}}$ be the covariant derivative operators on M^o and $\mathbb{C}^{n+1}$ respectively, as in (1.15). $D^{\mathbb{C}}_v$ is simply the directional derivative in the direction v. The relationship between D^M and $D^{\mathbb{C}}$ is given by the Gauss equation (see Hicks [8, p. 75]), which implies:

(7.7.1) Let $X \in M^o$, $\psi \in T_X M$; then for any vector field $\underline{v}$ tangent to M^o,

$$D^M_{\underline{\psi}}\underline{v} = (\text{grad } f)^{\perp}(D^{\mathbb{C}}_{\underline{\psi}}\underline{v}).$$

LEMMA 7.8. Let $X \in M^o$, $v \in T_X\mathbb{C}^{n+1}$, $\psi \in T_X M$. Then:

(1) $|D_v^{\mathbb{C}}(\underline{\text{grad}}\ f)|(X) = |v|O(|X|^{p-2})$;

(2) $|\underline{\langle}\psi,\ D_{\underline{\rho}}^{\mathbb{C}}(\underline{\text{grad}}\ f)\underline{\rangle}|(X) = |\psi|O(|X|^{p-1})$;

(3) $|(\underline{\text{grad}}\ f)^{\perp}(D_{\underline{\rho}}^{\mathbb{C}}\ \underline{\text{grad}}\ f)|(X) = O(|X|^{p-1})$.

Proof of (1). Let v have components $v_1,\ldots,v_{n+1}$ in the standard $\mathbb{C}$-basis of $T_X\mathbb{C}^{n+1}$. Then in the $\mathbb{R}$-basis $(\underline{\partial}_1,\ldots,\underline{\partial}_{n+1},\overline{\underline{\partial}}_1,\ldots,\overline{\underline{\partial}}_{n+1})$ of $T_X\mathbb{C}^{n+1}$, v has components

$$v = \sum_{1}^{n+1}(v_i\underline{\partial}_i + \overline{v}_i\overline{\underline{\partial}}_i)(X).$$

Therefore, since $\underline{\text{grad}}$ f is anti-holomorphic,

$$(7.8.4)\qquad D_v^{\mathbb{C}}(\underline{\text{grad}}\ f)(X) = \sum_{1}^{n+1}\overline{v}_i\overline{\underline{\partial}}_i(\underline{\text{grad}}\ f)$$
$$= \overline{D^2 f\cdot v},$$

by (7.1.4). Thus

$$|D_v^{\mathbb{C}}(\underline{\text{grad}}\ f)|(X) \le \|D^2 f(X)\|\,|v|.$$

Since f has order p at P,

$$\|D^2 f(X)\| = O(|X|^{p-2}).$$

Now (1) follows.

Proof of (2). By (7.8.4),

$$(D_{\underline{R}}^{\mathbb{C}}\ \underline{\text{grad}}\ f)(X) = (\overline{D^2 f\cdot\underline{R}})(X).$$

By (7.1.2), the jth component of this vector is

$$\sum_i \overline{\underline{\partial}_i(\underline{\partial}_j f)\cdot X_i} = (p-1)\overline{\underline{\partial}_j f(X)} + (\text{terms of order} \ge p),$$

by Euler's theorem. Hence, by (7.1.3),

$$(D^{\mathbb{C}}_{\underline{R}}\ \underline{\text{grad}}\ f)(X) = (p-1)\ \underline{\text{grad}}\ f(X) + (\text{terms of order} \geq p).$$

That is,

$$(7.8.5)\qquad |D^{\mathbb{C}}_{\underline{R}}(\underline{\text{grad}}\ f) - (p-1)\underline{\text{grad}}\ f|(X) = O(|X|^p).$$

Now

$$D^{\mathbb{C}}_{\underline{R}^{\perp}} - D^{\mathbb{C}}_{\underline{R}} = D^{\mathbb{C}}_{\underline{R}^{\perp}-\underline{R}}\ ;$$

and since, by (7.2.4),

$$|\underline{R}^{\perp} - \underline{R}|(X) = O(|X|^2)$$

(1) gives

$$|D^{\mathbb{C}}_{\underline{R}^{\perp}}(\underline{\text{grad}}\ f) - D^{\mathbb{C}}_{\underline{R}}(\underline{\text{grad}}\ f)|(X) = O(|X|^p).$$

Combining this with (7.8.5) gives

$$(7.8.6)\qquad |D^{\mathbb{C}}_{\underline{R}^{\perp}}(\underline{\text{grad}}\ f) - (p-1)\underline{\text{grad}}\ f|(X) = O(|X|^p).$$

Since $\underline{\langle}\psi,\underline{\text{grad}}\ f\underline{\rangle}(X) = 0$,

$$\begin{aligned}(7.8.7)\qquad |\underline{\langle}\psi,D^{\mathbb{C}}_{\underline{R}^{\perp}}\ \underline{\text{grad}}\ f\underline{\rangle}|(X) &= |\underline{\langle}\psi,D^{\mathbb{C}}_{\underline{R}^{\perp}}(\underline{\text{grad}}\ f) - (p-1)\underline{\text{grad}}\ f\underline{\rangle}|(X)\\ &\leq |\psi|\,|D^{\mathbb{C}}_{\underline{R}^{\perp}}(\underline{\text{grad}}\ f) - (p-1)\underline{\text{grad}}\ f|(X)\\ &= |\psi|O(|X|^p),\end{aligned}$$

by (7.8.6). By (7.3.3),

$$\begin{aligned}|\underline{\langle}\psi,D^{\mathbb{C}}_{\underline{\rho}}\ \underline{\text{grad}}\ f\underline{\rangle}|(X) &= (|\underline{R}|\,|\underline{R}^{\perp}|^{-2}\underline{\langle}\psi,D^{\mathbb{C}}_{\underline{R}^{\perp}}\ \underline{\text{grad}}\ f\underline{\rangle})(X)\\ &= |X|^{-1}(1 + O(|X|))|\psi|O(|X|^p),\end{aligned}$$

by (7.1.2), Lemma 7.2.3 and (7.8.7),

$$= |\psi| O(|X|^{p-1});$$

which proves (2).

Proof of (3). For any vector $v \in T_X\mathbb{C}^{n+1}$

$$|(\underline{\text{grad}}\ f)^{\perp}v| = \sup\{|\langle \psi, v\rangle| \text{ for } \psi \in T_X M \text{ with } |\psi| = 1\}.$$

So (3) follows from (2). q.e.d.

(7.9) Proof of Theorem 7.6, concluded. It remains to prove V and VII of (1.6.3).

Proof of V. Let $X \in M^o$, $\psi \in T_X M$. Then

$$D_\psi^M\, \underline{R}^{\perp} = (\underline{\text{grad}}\ f)^{\perp}(D_\psi^{\mathbb{C}}\, \underline{R}^{\perp}),$$

by (7.7.1). By (7.1.2) and (7.1.12),

$$\underline{R}^{\perp} = \underline{R} - |\underline{\text{grad}}\ f|^{-2}\langle \underline{R}, \underline{\text{grad}}\ f\rangle \underline{\text{grad}}\ f.$$

By (7.7) and (7.1.2),

$$D_\psi^{\mathbb{C}}\underline{R} = \partial_\psi \underline{R} = \psi.$$

Hence

$$(7.9.1) \qquad D_\psi^M \underline{R}^{\perp} = \psi - |\underline{\text{grad}}\ f|^{-2}\langle \underline{R}, \underline{\text{grad}}\ f\rangle (\underline{\text{grad}}\ f)^{\perp}(D_\psi^{\mathbb{C}}\ \underline{\text{grad}}\ f)$$
$$= \psi + \chi,$$

say, where, by Lemmas 7.2.1, 7.2.2 and 7.8.1,

$$(7.9.2) \qquad |\chi| = |\psi| O(|X|).$$

When applied to functions, the covariant derivative D^M operates simply as the directional derivative. So

$$(7.9.3) \qquad D_\psi^M(|\underline{R}|\,|\underline{R}^{\perp}|^{-2}) = |\underline{R}|^{-1}|\underline{R}^{\perp}|^{-2}\langle \psi, \underline{R}\rangle - 2|\underline{R}|\,|\underline{R}^{\perp}|^{-4}\langle D_\psi^M \underline{R}^{\perp}, \underline{R}^{\perp}\rangle.$$

Since $\langle \psi, \underline{\text{grad}}\ f\rangle = 0$, (7.1.12) shows that

$$\langle \psi,\underline{R}\rangle = \langle \psi,\underline{R}^{\perp}\rangle.$$

Making this substitution in (7.9.3) yields

$$(7.9.4)\quad (D^M_\psi(|\underline{R}|\,|\underline{R}^{\perp}|^{-2}))\underline{R}^{\perp} = |\underline{R}|^{-1}(\underline{R}^{\perp})^{\parallel}(\psi) - 2|\underline{R}|\,|\underline{R}^{\perp}|^{-2}(\underline{R}^{\perp})^{\parallel}(D^M_\psi\underline{R}^{\perp})$$
$$= |\underline{R}|^{-1}\underline{\rho}^{\parallel}(\psi) - 2|\underline{R}|\,|\underline{R}^{\perp}|^{-2}\underline{\rho}^{\parallel}(\psi + \chi),$$

using (7.9.1) and the fact that $\underline{\rho}$ is a real scalar multiple of $\underline{R}^{\perp}$ by (7.3.3).

By (7.3.3) again,

$$D^M_\psi\underline{\rho} = (D^M_\psi(|\underline{R}|\,|\underline{R}^{\perp}|^{-2}))\underline{R}^{\perp} + |\underline{R}|\,|\underline{R}^{\perp}|^{-2}D^M_\psi(\underline{R}^{\perp})$$
$$= |\underline{R}|^{-1}\underline{\rho}^{\parallel}(\psi) - 2|\underline{R}|\,|\underline{R}^{\perp}|^{-2}\underline{\rho}^{\parallel}(\psi+\chi) + |\underline{R}|\,|\underline{R}^{\perp}|^{-2}(\psi+\chi)$$

by (7.9.1) and (7.9.4). Now by (7.1.2),

$$|\underline{R}|^{-1}(X) = |X|^{-1};$$

and from this and Lemma 7.2.3,

$$|\underline{R}|\,|\underline{R}^{\perp}|^{-2}(X) = |X|^{-1}(1 + O(|X|)).$$

It follows that

$$(D^M_\psi\underline{\rho})(X) = |X|^{-1}\underline{\rho}^{\perp}(\psi) + O(|X|^0)(\underline{\rho}^{\perp}(\psi) - \underline{\rho}^{\parallel}(\psi))$$
$$+ O(|X|^{-1})(\underline{\rho}^{\perp}(\chi) - \underline{\rho}^{\parallel}(\chi)).$$

But

$$|\underline{\rho}^{\perp}(\psi) - \underline{\rho}^{\parallel}(\psi)| \le 2|\psi|;\ \text{and}$$
$$|\underline{\rho}^{\perp}(\chi) - \underline{\rho}^{\parallel}(\chi)| \le 2|\chi|$$
$$= |\psi|O(|X|),$$

by (7.9.2). Therefore

$$|(D^M_\psi\underline{\rho})(X) - |X|^{-1}\underline{\rho}^{\perp}(\psi)| = O(|X|^0).$$

Substitute $X = h(Z,t)$; then $|X| = t$ by (7.5.4), and V now follows with

$k = 1/2$.

Proof of VII. Let $\underline{v}$, $\underline{w}$ and $\underline{z}$ be vector fields tangent to M^o and let $X \in M^o$. By (7.7.1),

$$D^M_{\underline{w}}\underline{z} = (\underline{\text{grad}}\ f)^{\perp}(D^{\mathbb{C}}_{\underline{w}}\underline{z})$$
$$= D^{\mathbb{C}}_{\underline{w}}\underline{z} - |\underline{\text{grad}}\ f|^{-2}\langle D^{\mathbb{C}}_{\underline{w}}\underline{z}, \underline{\text{grad}}\ f\rangle \underline{\text{grad}}\ f,$$

by (1.13.6). Since $\underline{z}$ is tangent to M^o, $\langle \underline{z}, \underline{\text{grad}}\ f\rangle \equiv 0$. Hence

$$\langle D^{\mathbb{C}}_{\underline{w}}\underline{z}, \underline{\text{grad}}\ f\rangle = -\langle \underline{z}, D^{\mathbb{C}}_{\underline{w}}\ \underline{\text{grad}}\ f\rangle.$$

Thus

$$D^M_{\underline{w}}\underline{z} = D^{\mathbb{C}}_{\underline{w}}\underline{z} - |\underline{\text{grad}}\ f|^{-2}\langle \underline{z}, D^{\mathbb{C}}_{\underline{w}}\ \underline{\text{grad}}\ f\rangle \underline{\text{grad}}\ f,$$

Now by (7.7.1) again,

$$(7.9.5)\quad D^M_{\underline{v}}D^M_{\underline{w}}\underline{z} = (\underline{\text{grad}}\ f)^{\perp}(D^{\mathbb{C}}_{\underline{v}}D^M_{\underline{w}}\underline{z})$$
$$= (\underline{\text{grad}}\ f)^{\perp}(D^{\mathbb{C}}_{\underline{v}}D^{\mathbb{C}}_{\underline{w}}\underline{z})$$
$$- |\underline{\text{grad}}\ f|^{-2}\langle \underline{z}, D^{\mathbb{C}}_{\underline{w}}\ \underline{\text{grad}}\ f\rangle(\underline{\text{grad}}\ f)^{\perp}(D^{\mathbb{C}}_{\underline{w}}\ \underline{\text{grad}}\ f).$$

Using (7.7.1) and (7.9.5) in (1.15.4) results in:

$$R^M(\underline{v},\underline{w})\underline{z} = (D^M_{\underline{v}}D^M_{\underline{w}} - D^M_{\underline{w}}D^M_{\underline{v}} - D^M_{[\underline{v},\underline{w}]})\underline{z}$$
$$= (\underline{\text{grad}}\ f)^{\perp}(D^{\mathbb{C}}_{\underline{v}}D^{\mathbb{C}}_{\underline{w}} - D^{\mathbb{C}}_{\underline{w}}D^{\mathbb{C}}_{\underline{v}} - D^{\mathbb{C}}_{[\underline{v},\underline{w}]})\underline{z}$$
$$+ |\underline{\text{grad}}\ f|^{-2}\{\langle \underline{z}, D^{\mathbb{C}}_{\underline{v}}\ \underline{\text{grad}}\ f\rangle(\underline{\text{grad}}\ f)^{\perp}(D^{\mathbb{C}}_{\underline{w}}\ \underline{\text{grad}}\ f)$$
$$- \langle \underline{z}, D^{\mathbb{C}}_{\underline{w}}\ \underline{\text{grad}}\ f\rangle(\underline{\text{grad}}\ f)^{\perp}(D^{\mathbb{C}}_{\underline{v}}\ \underline{\text{grad}}\ f)\}.$$

The first term on the right-hand side is

$$(\underline{\text{grad}}\ f)^{\perp}R^{\mathbb{C}}(\underline{v},\underline{w})\underline{z} = 0$$

because $\mathbb{C}^{n+1}$ is flat, so its curvature tensor $R^{\mathbb{C}}$ is identically zero. It follows that

$$(7.9.6)\quad R^M(\underline{v},\underline{w})\underline{z} = |\underline{\text{grad}}\ f|^{-2}\{\langle \underline{z}, D^{\mathbb{C}}_{\underline{v}}\ \underline{\text{grad}}\ f\rangle(\underline{\text{grad}}\ f)^{\perp}\!\!-(D^{\mathbb{C}}_{\underline{w}}\ \underline{\text{grad}}\ f)$$

$$- \langle \underline{z}, D^{\mathbb{C}}_{\underline{w}}\ \underline{\text{grad}}\ f\rangle(\underline{\text{grad}}\ f)^{\perp}\!\!-(D^{\mathbb{C}}_{\underline{v}}\ \underline{\text{grad}}\ f)\}.$$

Therefore

$$(7.9.7)\quad |R^M(\underline{v},\underline{w})\underline{z}|(X) \le 2(|\underline{\text{grad}}\ f|^{-2}|\underline{z}||D^{\mathbb{C}}_{\underline{v}}\ \underline{\text{grad}}\ f||D^{\mathbb{C}}_{\underline{w}}\ \underline{\text{grad}}\ f|)(X)$$

$$= O(|X|^{-2})(|\underline{v}||\underline{w}||\underline{z}|)(X),$$

by Lemmas 7.2.1 and 7.8.1. Hence

$$\| R^M(\underline{v},\underline{w})\|(X) = O(|X|^{-2})(|\underline{v}||\underline{w}|)(X).$$

Substitute $X = h(Z,t)$; then $|X| = t$ by (7.5.4), and VII(i) follows.

Now let $\underline{v} = \underline{\rho}$. I apply Lemmas 7.8.2 and 7.8.3 respectively to the terms

$$\langle \underline{z}, D^{\mathbb{C}}_{\underline{\rho}}\ \underline{\text{grad}}\ f\rangle \qquad \text{and}$$

$$(\underline{\text{grad}}\ f)^{\perp}\!\!-(D^{\mathbb{C}}_{\underline{\rho}}\ \underline{\text{grad}}\ f)$$

on the right-hand side of (7.9.6). This improves (7.9.7) to

$$(7.9.8)\quad |R^M(\underline{\rho},\underline{w})\underline{z}|(X) = O(|X|^{-1})(|\underline{\rho}||\underline{w}||\underline{z}|)(X)$$

$$= O(|X|^{-1})(|\underline{w}||\underline{z}|)(X),$$

by (7.6.4) (in which $X = h(Z,t)$). Hence

$$\| R^M(\underline{\rho},\underline{w})\|(X) = O(|X|^{-1})|\underline{w}(X)|;$$

as above, this proves VII(ii) with $k = 1$ this time.

Finally if $\underline{v} = \underline{z} = \underline{\rho}$, then (7.9.8) and (7.6.3) again give

$$|R^M(\underline{\rho},\underline{w})\underline{\rho}|(X) = O(|X|^{-1})|\underline{w}(X)|;$$

again, this implies VII(iii) with $k = 1/2$.

This completes the proof of Theorem 7.6, using $k = 1/2$ in V and VII of (1.6.3). q.e.d.

THEOREM 7.10. Let P be an unbranched isolated singular point of a complex analytic hypersurface M, and let T be the Whitney tangent cone of M at P. Then there is a homeomorphism $T_P M \approx T$ under which P corresponds to P.

Proof. Let M have equation $f(X) = 0$ in some neighbourhood U of P in $\mathbb{C}^{n+1}$ where $f = g+h$ as in (7.1), so that g is the initial homogeneous polynomial of f. Define $F\colon U \times \mathbb{C} \to \mathbb{C}$ by

$$(7.10.1) \quad F(X,z) = g(X) + zh(X).$$

Let $\underline{t}$ be as in (7.3) and let Δ be the polycylinder

$$\Delta = \{|X| \leq \underline{t},\ |z| \leq 1\} \subseteq U \times \mathbb{C}.$$

The proof of Lemma 7.2.1 shows that on Δ, $\underline{\text{grad}}\ F(X,z) = \bar{0}$ if and only if $X = P$. Set

$$\mathcal{X} = F^{-1}(0) \cap \Delta,\ \mathcal{Y} = \{(0,z) \text{ for } |z| \leq 1\},$$

and define $\pi\colon \mathcal{X} \to \mathcal{Y}$ by

$$\pi(X,z) = (0,z).$$

For each $z \in \mathbb{C}$ such that $|z| \leq 1$, set

$$(7.10.2) \quad \mathcal{X}_z = \pi^{-1}(0,z),\ F_z = F \restriction \mathcal{X}_z.$$

Each $\mathcal{X}_z$, regarded as a hypersurface in $\mathbb{C} \times z$, has an isolated singularity at (P,z); this applies also to the "exceptional" fibre $\mathcal{X}_0$ by (7.1.10). Let $\mu(\mathcal{X}_z)$ be the Milnor number of $\mathcal{X}_z$ (see Milnor [9, p. 59]), namely the degree of the map $\lambda_z\colon\ \$(\underline{t}) \to (\mathbb{C}^{n+1})^o$ defined by

$$\lambda_z(X) = (\underline{\partial}_1 F_z, \ldots, \underline{\partial}_{n+1} F_z)(X,z)$$

(see (7.1.1)). Thus

$$\lambda_z(X) = (\underline{\partial}_1 g, \ldots, \underline{\partial}_{n+1} g)(X) + z(\underline{\partial}_1 h, \ldots, \underline{\partial}_{n+1} h)(X).$$

Fix z, and define a homotopy $(\lambda_z)_t$ between λ_0 and λ_z by $(\lambda_z)_t = \lambda_{tz}$. Since λ_0 and λ_z are homotopic through maps $\$(\underline{t}) \to (\mathbb{C}^{n+1})^o$, they have the same degree. Thus

(7.10.3) $\mu(X_z)$ is constant, for $|z| \leq 1$.

Teissier [15, p. 300] defines a sequence of numbers for each z,

(7.10.4) $\mu^*(X_z) = (\mu^{n+1}(X_z),\ldots,\mu^0(X_z))$,

as follows. He shows that given a complex analytic hypersurface $X' \subseteq \mathbf{C}^{n+1}$ with an isolated singular point P', then there is a Zariski-open set $\mathcal{U}^r(X')$ in the Grassmannian of (complex) r-planes through P' in $\mathbf{C}^{n+1}$ such that for every $H \in \mathcal{U}^r(X')$, $X'(H) = X' \cap H$ is a complex analytic hypersurface in H with an isolated singularity at P'. Moreover

$$\mu^r(X') = \mu(X'(H))$$

is independent of the choice of H. Thus the sequence (7.10.4) is defined.

Returning to (7.10.2), let T_z be the Whitney tangent cone of X_z at (P,z). From (7.10.1), $T_z = T \times z$. Hence for all $H \in \mathcal{U}^r(T)$, $T_z(H)$ has an isolated singularity in H at (P,z); and it follows from Lemma 7.2.1 that the same is true for $X_z(H)$. Thus for such H, (P,z) is an unbranched isolated singularity of $X_z(H)$ in H. Applying (7.10.3) to the family $X_z(H)$ shows that in (7.10.4),

$$\mu^*(X_z) \text{ is constant, for } |z| \leq 1.$$

It now follows from a theorem of Teissier [15, p. 334] that the pair $(X,\mathcal{Y})$ satisfies Whitney's conditions (a) and (b) of [19, p. 228]). A result of Thom ([16, Corollaire 2.C.2]) implies that the pairs $(X_z,(P,z))$ are all homeomorphic. Since $(X_0,(P,0))$ and $(X_1,(P,1))$ can be identified with neighbourhoods of P in (T,P) and in (M,P) respectively, it follows that (T,P) and (M,P) are homeomorphic pairs near P. By Theorem 2.26.2, the pairs (M,P) and $(T_PM,\hat{P})$ are homeomorphic near P and $\hat{P}$. Therefore (T,P) and $(T_PM,\hat{P})$ are homeomorphic near P and $\hat{P}$. Since T and T_PM are cones, the theorem follows. q.e.d.

(7.10.5) Conjecture.. Let α be a path in M from P that is C^1 at P. Then α, regarded as a path in $\mathbb{C}^{n+1}$, is differentiable at 0, and so has a tangent vector $\alpha'(0) \in T_P\mathbb{C}^{n+1} = \mathbb{C}^{n+1}$. It is elementary to check that equivalent paths in M from P have the same tangent vector. It is also known (see Whitney [19, Theorem 5.8]) that $\alpha'(0)$ always lies in T, the Whitney tangent cone of M at P. Thus there is a natural map of pairs

$$\hat{\pi}: \quad (T_PM, \hat{P}) \to (T,P).$$

I conjecture that $\hat{\pi}$ is a homeomorphism and even an isometry with respect to the metric d^T on T_PM and the intrinsic metric on T (namely, that metric constructed in (2.4.1) from the Riemannian metric of (7.1.11) on T^o). Theorem 7.18 implies the truth of the conjecture when n = 1; and (7.27) is a more general conjecture.

One-Dimensional Hypersurfaces

Theorem 7.6 can be strengthened in case n = 1. For curves in $\mathbb{C}^2$, any singular point is isolated. Unbranched singular points are those at which only normal crossings occur; in this situation Theorem 7.6 is elementary, as will be explained in (7.11.5). However the hypothesis that singular points be unbranched is no longer needed when n = 1.

THEOREM 7.11. Let M be a complex analytic curve in $\mathbb{C}^2$ and let P be a singular point of M. Then P is an isolated conical singularity of M, and so M has an exponential map at P satisfying (1.1) -- (1.5).

Proof. The proof incorporates Lemmas 7.12, 7.14 and 7.15 and is concluded in (7.16).

(7.11.1) Notation. I write X = (x,y) for a point of $\mathbb{C}^2$ and $v = (v_x, v_y)$ for the complex components of a vector.

A holomorphic function f(x,y) defined in a neighbourhood of P is a Weierstrass polynomial in y if f can be expressed, in a neighbourhood of P, as

$$(7.11.2)\quad f(x,y) = y^p + \sum_0^{p-1} a_\ell(x)y^\ell,$$

where the a_ℓ are holomorphic in x. (See Whitney [20, Chapter 1, Definition 5G].) It is possible to find a unitary change of coordinates in $\mathbb{C}^2$ after which M has equation $f(x,y) = 0$ near P with f a Weierstrass polynomial in y; this follows from the Weierstrass preparation theorem (see for example Whitney [20, Chapter 1, Theorem 5I]).

The first step in proving Theorem 7.11 is to reduce it to the special case that M is topologically unibranch at P in the sense of Mumford [10, p. 43]; namely:

(7.11.3) the germ of f is irreducible as a Weierstrass polynomial in y.

In the general case, say

$$f = f_1^{m_1} \cdots f_r^{m_r}$$

is the factoring (near P) of f into irreducible Weierstrass polynomials in y, in which the f_i are distinct. Set

$$M_i = \{f_i = 0\}.$$

Then $M = \cup M_i$ and the M_i meet pairwise at a finite set of points. In a sufficiently small neighbourhood of P, I may therefore assume

$$(7.11.4)\quad M = \bigcup_1^r M_i, \text{ where}$$

$$M_i \cap M_j = \{P\}, \text{ for } i \neq j.$$

Assume for the moment that Theorem 7.11 is valid under the additional hypothesis (7.11.3). Then for each $i = 1,\ldots,r$ there exist manifolds L_i and charts $h_i\colon cL_i \to M$ satisfying I -- VII of (1.6.3). (The proof of Lemma 7.12 will show that each L_i is a circle.) Let L be the disjoint union of the L_i, and define $h\colon cL \to M$ by the rule

$$h \upharpoonright cL_i = h_i.$$

Then I -- VII of (1.6.3) hold for h, so the theorem is proved in general.

(7.11.5) Remark. If M is unbranched at P, that is, if M has only normal crossings at P, then each M_i in (7.11.4) is non-singular at P. By (1.7.2) charts h_i for M_i at P exist; and now the argument above establishes Theorem 7.6 in case n = 1.

It remains to prove Theorem 7.11 under the additional assumption (7.11.3).

LEMMA 7.12. Let f be a Weierstrass polynomial in y on a neighbourhood of P in $\mathbb{C}^2$, whose germ at P is irreducible in the sense of (7.11.3). Let g be the initial homogeneous polynomial of f. Then g has the form

$$g(x,y) = (ay - bx)^p,$$

for some $a,b \in \mathbb{C}$ not both 0, and some integer $p \geq 1$.

Proof. Suppose g could be factored as

$$g = \prod_1^s (a_i y - b_i x)^{p_i}.$$

where $s > 1$ and where the lines

$$L_i = \{a_i y - b_i x = 0\}$$

are distinct for $i = 1,\dots,s$. Then, for $i \neq j$, $L_i \cap L_j = \{P\}$. Let $\varepsilon > 0$, and for each i, set

$$U_i = \{(x,y) \text{ such that } |a_i y - b_i x| \leq \varepsilon |(x,y)|.$$

Then, for sufficiently small ε,

$$U_i \cap U_j = \{P\} \text{ for } i \neq j.$$

Now by Whitney [20, Chapter 7, Lemma 3G] the domain of f can be chosen so small that

$$M \subseteq \bigcup_1^s U_i.$$

The <u>multifunction</u> F of f is defined in [20, Chapter 1, section 9] by

$$F(x) = \{(x,y) \text{ such that } f(x,y) = 0, \text{ each point counted with its multiplicity}\}.$$

Set

$$F_i(x) = \{(x,y) \text{ such that } (x,y) \in M \cap U_i, \text{ each point being counted with its multiplicity in } F(x)\}.$$

Then the F_i determine a <u>splitting</u> of F:

$$F = F_1 \oplus \ldots \oplus F_s,$$

in the terminology of [20, Appendix V, Definition 10E]. Then by [20, Chapter 1, Theorem 9G], there is a corresponding factoring

(7.12.1) $f = f_1 \ldots f_s$

of f into Weierstrass polynomials in y, such that each f_i has F_i for multifunction. It follows that each f_i is non-trivial, so that (7.12.1) contradicts (7.11.3). q.e.d.

COROLLARY 7.12.2. Let f satisfy (7.11.2) and (7.11.3) as in the hypotheses of Lemma 7.12. Then the coordinate system in $\mathbb{C}^2$ may be adjusted by a unitary transformation so that near P M has equation

$$f(x,y) = \xi(x) - y^p = 0,$$

where

$$\xi(x) = ax^q + \sum_{q+1}^{\infty} a_i x^i,$$

with $q > p$ and $a \neq 0$.

<u>Proof</u>. Choose the x-axis to be the tangent line of M at P, given by Lemma 7.12. By Whitney [20, Chapter 1, Theorem 10A], M has an equation near P of the form $f(x,y) = \xi(x) - y^p = 0$. Since the tangent cone of M at P is the x-axis with some multiplicity, and is the zero-set of the initial homo-

geneous polynomial of f, it follows that ξ has order >p at P. Now the corollary follows. q.e.d.

(7.13) <u>Proof of Theorem 7.11, continued</u> under the additional assumption (7.11.3). Assume M is defined near P by the equation $f(x,y) = 0$, where f is as in Corollary 7.12.2. Observe that when $X = (x,y) \in M^o$, then:

(7.13.1) $|y| = |a|^{q/p}|x|^{q/p}(1 + O(|x|^{1/p})) = O(|x|^{q/p})$;

(7.13.2) $|y|^{-1} = |a|^{-q/p}|x|^{-q/p}(1 + O(|x|^{1/p})) = O(|x|^{-q/p})$;

(7.13.3) $|X| = |x|(1 + O(|x|^{2(q/p-1)}))$.

The proof of Theorem 7.11 is henceforth analogous to that of Theorem 7.6 but uses the assertions of Lemmas 7.14 and 7.15 instead of the correspondingly numbered assertions of Lemmas 7.2 and 7.8 respectively.

LEMMA 7.14. Let the notation be as in (7.13). Then:

(1) $|\underline{\text{grad}}\ f|^{-1}(X) = O(|X|^{-q+q/p})$;

(2) $|\langle \underline{R}, \underline{\text{grad}}\ f\rangle|(X) = O(|X|^q)$;

(3) $|\underline{R}^{\perp}(X)| = |X|(1 + O(|X|^{2(q/p-1)}))$.

<u>Proof of (1)</u>. By (7.1.3),

$$\underline{\text{grad}}\ f(X) = (\overline{\xi'(x)}, -p\,\bar{y}^{p-1}).$$

By Corollary 7.12.2, ξ' has order q-1 in x; so

$$|\overline{\xi'(x)}| = O(|x|^{q-1}) = O(|X|^{q-1}),$$

by (7.13.3). By (7.13.1),

$$|-p\,\bar{y}^{p-1}| = c|x|^{q(p-1)/p}(1 + O(|x|^{1/p})) \text{ for some } c \neq 0$$
$$= c|X|^{q-q/p}(1 + O(|X|^{1/p})),$$

by (7.13.3) again. The order of $|\underline{\text{grad}}\ f|(X)$ is the maximum of the orders of $|\overline{\xi'(x)}|$ and $|-p\ \bar{y}^{p-1}|$; and since $q-q/p < q-1$,

$$|\underline{\text{grad}}\ f|(X) = c|X|^{q-q/p}(1 + O(|X|^{1/p})),$$

from which (1) follows.

Proof of (2). By (7.1.2) and Corollary 7.12.2,

$$\begin{aligned}\langle \underline{R}, \underline{\text{grad}}\ f\rangle(X) &= x\xi'(x) - p\ y^p \\ &= (q\xi(x) + O(x^{q+1})) - p\xi(x) \text{ on } M^o \\ &= (q - p)\xi(x) + O(x^{q+1}).\end{aligned}$$

Since ξ has order q in x,

$$\begin{aligned}|\langle \underline{R}, \underline{\text{grad}}\ f\rangle|(X) &= O(|x|^q) \\ &= O(|X|^q),\end{aligned}$$

by (7.13.3). This proves (2).

Proof of (3). This is analogous to the proof of Lemma 7.2.3.

For future reference, the analogue of (7.2.4) is:

(7.14.4) $|\underline{R} - \underline{R}^{\perp}|(X) = O(|X|^{q/p})$. q.e.d.

LEMMA 7.15. Let the notation be as in (7.13). Then for $X \in M^o$, $v \in T_X\mathbb{C}^{n+1}$ and $\psi \in T_XM^o$:

(1) $|D_v^{\mathbb{C}}\ \underline{\text{grad}}\ f|(X) = |v|O(|X|^{q-2q/p})$;

(1a) $|\langle \underline{\psi}, D_v^{\mathbb{C}}\ \underline{\text{grad}}\ f\rangle|(X) = |\psi||v|O(|X|^{q-1-q/p})$;

(1b) $|(\underline{\text{grad}}\ f)^{\perp}(D_v^{\mathbb{C}}\ \underline{\text{grad}}\ f)|(X) = |v|O(|X|^{q-1-q/p})$;

(2) $|\langle \underline{\psi}, D_{\underline{\rho}}^{\mathbb{C}}\ \underline{\text{grad}}\ f\rangle|(X) = |\psi|O(|X|^{q-2})$;

(3) $|(\underline{\text{grad}}\ f)^{\perp}(D_{\underline{\rho}}^{\mathbb{C}}\ \underline{\text{grad}}\ f)|(X) = O(|X|^{q-2})$.

Proof of (1). As in the proof of Lemma 7.8.1, (7.8.4) still holds:

(7.15.4) $D_v^{\mathbb{C}}\ \underline{\text{grad}}\ f = \overline{D^2 f\cdot v}$;

so

(7.15.5) $|D_v^{\mathbb{C}}\ \underline{\text{grad}}\ f| \leq |v| \| D^2 f\|$.

Now $D^2 f$ is a 2×2 matrix whose non-zero entries are:

$$h''(x) = O(x^{q-2}),$$
$$p(p-1)y^{p-2} = O(x^{q-2q/p}),$$

by (7.13.1). Since $q-2q/p < q-2$, it follows that

$$\| D^2 f(X)\| = O(|x|^{q-2q/p})$$
$$= O(|X|^{q-2q/p}),$$

by (7.13.3). Now (1) follows from (7.15.5).

<u>Proof of (1a)</u>. By (7.15.4), and using the notation of (1.14.8),

$$\underline{\langle}\psi, D_v^{\mathbb{C}}\ \underline{\text{grad}}\ f\underline{\rangle} = \psi \cdot D^2 f \cdot v$$
$$= v \cdot D^2 f \cdot \psi$$
$$= \underline{\langle} v, D_\psi^{\mathbb{C}}\ \underline{\text{grad}}\ f\underline{\rangle}.$$

Therefore

$$|\underline{\langle}\psi, D_v^{\mathbb{C}}\ \underline{\text{grad}}\ f\underline{\rangle}| \leq |v||D_\psi^{\mathbb{C}}\ \underline{\text{grad}}\ f|;$$

so it suffices to prove:

(7.15.6) $|D_\psi^{\mathbb{C}}\ \underline{\text{grad}}\ f|(X) = |\psi| O(|X|^{q-1-q/p})$.

Now by (7.15.4).

(7.15.7) $(D_\psi^{\mathbb{C}}\ \underline{\text{grad}}\ f)(X) = (\overline{\psi_x \xi''(x)}, -p(p-1)\overline{\psi_y y^{p-2}})$

Since ψ is tangent to M^o,

(7.15.8) $0 = \underline{\langle}\psi, \underline{\text{grad}}\ f(X)\underline{\rangle}$

$$= \psi_x \xi'(x) - p\psi_y y^{p-1}.$$

By (7.13.2), y is nowhere zero on M^o; so I may use (7.15.8) to substitute for y in (7.15.7), to obtain

$$(7.15.9) \quad (D_\psi^{\mathbb{C}} \underline{\text{grad}}\ f)(X) = \overline{\psi_x}(\overline{\xi''(x)}, -(p-1)\overline{\xi'(x)}/\bar{y}).$$

Now ξ'' and ξ' are of orders q-2 and q-1, respectively, in x, by Corollary 7.12.2. Hence

$$|\xi''(x)| = O(|x|^{q-2});$$

and, in view of (7.13.2),

$$|\xi'(x)/y| = O(|x|^{q-1-q/p}).$$

Hence (7.15.9) gives

$$\begin{aligned} |D_\psi^C \underline{\text{grad}}\ f|(X) &= |\psi_x| O(|x|^{q-1-q/p}) \\ &\leq |\psi| O(|x|^{q-1-q/p}), \end{aligned}$$

by (7.13.3). This proves (7.15.6) and with it, (1a).

Proof of (1b). This is analogous to the proof of Lemma 7.8.3.

Proof of (2). By (7.15.4),

$$\begin{aligned} (7.15.10) \quad (D_{\underline{R}}^{\mathbb{C}} \underline{\text{grad}}\ f)(X) &= (\overline{x\xi''(x)}, p(p-1)\bar{y}^{p-1}) \\ &= ((q-1)\overline{\xi'(x)} + O(\bar{x}^q), (p-1)p\ \bar{y}^{p-1}) \\ &= (p-1)\underline{\text{grad}}\ f(X) + \eta, \end{aligned}$$

say, where

$$\eta = ((q-p)\overline{\xi'(x)} + O(\bar{x}^q), 0),$$

so that

$$\begin{aligned} (7.15.11) \quad |\eta| &= O(|x|^{q-1}) \\ &= O(|X|^{q-1}), \end{aligned}$$

by (7.13.3). Now

(7.15.12) $$\underline{\langle}\psi,D^{\mathbb{C}}_{\underline{R}^{\perp}}\ \underline{\text{grad}}\ f\underline{\rangle}(X) = \underline{\langle}\psi,D^{\mathbb{C}}_{\underline{R}}\ \underline{\text{grad}}\ f\underline{\rangle}(X) + \underline{\langle}\psi,D^{\mathbb{C}}_{\underline{R}^{\perp}-\underline{R}}\ \underline{\text{grad}}\ f\underline{\rangle}(X).$$

I estimate the orders of magnitude of the two terms on the right-hand side of (7.15.12) separately.

$$\begin{aligned} |\underline{\langle}\psi,D^{\mathbb{C}}_{\underline{R}}\ \underline{\text{grad}}\ f\underline{\rangle}|(X) &= |\underline{\langle}\psi,\eta\underline{\rangle}|, \qquad \text{by (7.15.10)}\\ &\le |\psi||\eta| \\ &= |\psi|O(|X|^{q-1}), \end{aligned}$$

by (7.15.11).

$$\begin{aligned} |\underline{\langle}\psi,D^{\mathbb{C}}_{\underline{R}^{\perp}-\underline{R}}\ \underline{\text{grad}}\ f\underline{\rangle}|(X) &\le |\psi||\underline{R}^{\perp}-\underline{R}|(X)O(|X|^{q-1-q/p}), \text{ by (1a)}\\ &\le |\psi|O(|X|^{q-1}), \end{aligned}$$

by (7.14.4). So (7.15.12) gives the estimate

(7.15.13) $$|\underline{\langle}\psi,D^{\mathbb{C}}_{\underline{R}^{\perp}}\ \underline{\text{grad}}\ f\underline{\rangle}|(X) = |\psi|O(|X|^{q-1}).$$

By (7.3.3),

$$\begin{aligned} |\underline{\langle}\psi,D^{\mathbb{C}}_{\underline{\rho}}\ \underline{\text{grad}}\ f\underline{\rangle}|(X) &= (|\underline{R}||\underline{R}^{\perp}|^{-2}|\underline{\langle}\psi,D^{\mathbb{C}}_{\underline{R}^{\perp}}\ \underline{\text{grad}}\ f\underline{\rangle}|)(X)\\ &= |\psi||X|^{-1}(1 + O(|X|^{2(q/p-1)})O(|X|^{q-1}), \end{aligned}$$

by Lemma 7.14.3 and (7.15.13),

$$= |\psi|O(|X|^{q-2}),$$

which proves (2).

<u>Proof of (3)</u>. This is analogous to the proofs of (1b) and of Lemma 7.8.3.

q.e.d.

(7.16) <u>Proof of Theorem 7.11, concluded</u> under the additional hypothesis (7.11.3). Only the verification of V and VII of (1.6.3) needs any further comment. I claim that in these assertions one may use

(7.16.1) $k = (1/2)\min\{1, q/p-1\}$.

In the proof of V, let χ be defined as in (7.9.1); then the analogue of (7.9.2) is

$$|\chi| = |\psi| O(|X|^{q/p-1}).$$

Following the argument of (7.9), one obtains

$$\begin{aligned} |D^M_{\psi}\underline{\rho} - |X|^{-1}\underline{\rho}^{\perp}(\psi)|(X) &= O(|X|^0)|\psi| + O(|X|^{-1})|\chi| \\ &= |\psi| O(|X|^{-1+2k}). \end{aligned}$$

Now V follows with k as in (7.16.1).

To prove VII, I substitute in (7.9.6) the five following estimates:

(7.16.2) $|\underline{\text{grad}}\ f|^{-2}(X) = O(|X|^{2q+2q/p})$, by Lemma 7.14.1;

(7.16.3) $|\langle \underline{z}, D^{\mathbb{C}}_{\underline{v}}\ \underline{\text{grad}}\ f\rangle|(X) = (|\underline{z}||\underline{v}|)(X) O(|X|^{q-1-q/p})$, by Lemma 7.15.1a;

(7.16.4) $|(\underline{\text{grad}}\ f)^{\perp}(D^{\mathbb{C}}_{\underline{w}}\ \underline{\text{grad}}\ f)|(X) = |\underline{w}|(X) O(|X|^{q-1-q/p})$, by Lemma 7.15.1b;

(7.16.5) $|\langle \underline{z}, D^{\mathbb{C}}_{\underline{\rho}}\ \underline{\text{grad}}\ f\rangle|(X) = |\underline{z}|(X) O(|X|^{q-2})$, by Lemma 7.15.2;

(7.16.6) $|(\underline{\text{grad}}\ f)^{\perp}(D^{\mathbb{C}}_{\underline{\rho}}\ \underline{\text{grad}}\ f)|(X) = O(|X|^{q-2})$, by Lemma 7.15.3

In (7.16.3) and (7.16.4) the roles of $\underline{v}$ and $\underline{w}$ can be interchanged. Now VII(i) follows from (7.9.6) and the estimates (7.16.2), (7.16.3) and (7.16.4). To prove VII(ii) requires all of (7.16.2) -- (7.16.6); one obtains

$$\begin{aligned} |R^M(\underline{\rho}, \underline{w})\underline{z}|(X) &= (|\underline{w}||\underline{z}|)(X) O(|X|^{q/p-3}) \\ &= (|\underline{w}||\underline{z}|)(X) O(|X|^{-2+2k}), \end{aligned}$$

by (7.16.1). Thus VII(ii) holds even with 2k in place of k. Now, as in the proof of VII(iii) in Theorem 7.6, VII(iii) holds here with k as in (7.16.1). q.e.d.

In the case of curves in $\mathbb{C}^2$ it is possible to identify the topology and metric structure on $T_P M$. I shall first describe the reduction to the special case that M is topologically unibranch in the sense of (7.11.3). In that special case, Theorem 7.18 characterizes $T_P M$ in terms of the Whitney tangent cone of M at P, and Corollary 7.18.4 is an intrinsic characterization of $T_P M$.

Let M be a complex analytic curve in $\mathbb{C}^2$ and P a singular point of M. Then M can be expressed in the form (7.11.4), in which each M_i is topologically unibranch. It follows from the definition (2.4.1) of d^M that whenever $X \in M_i^{\,o}$ and $Y \in M_j^{\,o}$ with $i \neq j$, then

$$d^M(X,Y) = d^{M_i}(X,P) + d^{M_j}(P,Y).$$

The definitions in (2.25) of $T_P M$ and d^T now give the following decompositions of $T_P M$ and of d^T:

$$T_P M = \bigcup_1^r T_P M_i \;, \qquad \text{where}$$

$$T_P M_i \cap T_P M_j = \{\hat{P}\}, \quad \text{for} \quad i \neq j$$

Moreover d^T is determined from the intrinsic metrics d^{T_i} on $T_P M_i$ as follows. Given $\hat{\alpha},\hat{\beta} \in T_P M$, say $\hat{\alpha} \in T_P M_i$ and $\hat{\beta} \in T_P M_j$; then

$$d^T(\hat{\alpha},\hat{\beta}) = \begin{cases} d^{T_i}(\hat{\alpha},\hat{\beta}) \;, & \text{if} \quad i = j \\ \\ d^{T_i}(\hat{\alpha},\hat{P}) + d^{T_j}(\hat{P},\hat{\beta}), & \text{if} \quad i \neq j. \end{cases}$$

Thus in order to determine the metric space $T_P M$ it is enough to determine each $T_P M_i$; in other words, the problem is reduced to the case that M is topologically unibranch.

(7.17) <u>Notation</u>. Let M be topologically unibranch at P, and say M has equation $f = 0$, where $f(x,y) = \xi(x) - y^p$ as in Corollary 7.12.2. Let T be the Whitney tangent cone of M at P; since T is simply the x-axis, I shall identify T with $\mathbb{C}$ and write x instead of (x,0) for a point of T. Let

$$\pi: M \to T$$

be the orthogonal projection of M into T. Given a curve α in M which is C^1 at P, then α is C^1 at P regarded as a path in $\mathbb{C}^2$, and $\alpha'(0) \in T$. It follows from the definition (see (2.25)) that if α and β are equivalent, then $\alpha'(0) = \beta'(0)$. Hence π induces a map

$$\hat{\pi}: T_P M \to T$$

by the rule

(7.17.1) $\hat{\pi}(\hat{\alpha}) = \alpha'(0) = (\pi \circ \alpha)'(0)$,

where α is any representative of $\hat{\alpha}$.

THEOREM 7.18. Let M be a curve in $\mathbb{C}^2$, P a singular point of M at which M is topologically unibranch. Let the notation be as in (7.17). Then:

(1) $\hat{\pi}$ is a conical map and $|\hat{\pi}(\hat{\alpha})| = |\hat{\alpha}|$; in particular $\hat{\pi}^{-1}(P) = \{\hat{P}\}$;

(2) $\hat{\pi}: (T_P M)^o \to T^o$ is a local isometry: given $\hat{\alpha} \in (T_P M)^o$ and $\hat{\beta} \in B(\hat{\alpha}, T_P M; |\hat{\alpha}|/4)$, then

$$d^T(\hat{\alpha}, \hat{\beta}) = |\hat{\pi}(\hat{\alpha}) - \hat{\pi}(\hat{\beta})|;$$

(3) $\hat{\pi}: (T_P M)^o \to T^o$ is a p-fold covering map, where p is the multiplicity of M at P.

COROLLARY 7.18.4. Let M and P be as in Theorem 7.18, and let p be the multiplicity of M at P. Then $T_P M$ is metrically the infinite cone (in the sense of (1.16.6)) on a circle of length $2\pi p$.

Proof of Theorem 7.18. The proof incorporates Lemmas 7.19 and 7.20, and is completed in (7.21). From the form of $\xi(x)$ in Corollary 7.12.2 it follows that the restriction of π to

$$\pi^o: M^o \to T^o$$

is a p-fold covering map near P. Given $x \in T^o$, set

(7.18.5) $B_x = B(x, T; |x|)$;

then π^o is topologically trivial over B_X. Given $X = (x,y) \in M^o$, let

(7.18.6) U_X be the sheet of $(\pi^o)^{-1}B_X$ that contains X.

Let q be the order of ξ at P, and let k be as in (7.16.1).

LEMMA 7.19. Let $X = (x,y) \in M^o$ and $\psi \in T_X M^o$. Then

$$|\psi|(1 + O(|x|^{4k})) = |D\pi^o(X)\cdot\psi| \le |\psi|.$$

Proof. In terms of the notation of (7.11.1), ψ can be written as

$$\psi = (\psi_x \underline{\partial_x} + \psi_y \underline{\partial_y} + \overline{\psi_x \underline{\partial_x}} + \overline{\psi_y \underline{\partial_y}})(X).$$

Then

(7.19.1) $D\pi^o(X)\cdot\psi = (\psi_x \underline{\partial_x} + \overline{\psi_x \underline{\partial_x}})(X);$

so the inequality is immediate. To prove the estimate, first observe that

$$\begin{aligned} 0 &= \underline{\langle \psi, \underline{\text{grad}}\ f(X) \rangle} \\ &= \psi_x \xi'(x) - \psi_y p y^{p-1}. \end{aligned}$$

By (7.13.1), y is nowhere zero on M^o; hence

$$\begin{aligned} |\psi_y| &= |\psi_x|p|(\xi'(x)/\xi(x))y| \\ &= |\psi_x|O(|x|^{q/p-1}), \qquad \text{by (7.13.1) again} \\ &= |\psi_x|O(|x|^{2k}), \end{aligned}$$

by (7.16.1). It follows that

$$|\psi| = |\psi_x|(1 + O(|x|^{4k})).$$

Thus

$$|\psi_x| = |\psi|(1 + O(|x|^{4k})).$$

The same estimate holds for $|\overline{\psi_x}|$. Since $\psi_x \underline{\partial_x}$ and $\overline{\psi_x}\,\overline{\underline{\partial_x}}$ are real-orthogonal vectors, the lemma now follows from (7.19.1). q.e.d.

COROLLARY 7.19.2. Let β be a C^1 curve in M which is C^1 at P if it goes from, to or through P. Set

$$B = \sup\{|\pi\circ\beta(\sigma)| \text{ for } \sigma \in [0,1]\}.$$

Then

$$L(\beta)(1 + O(B^{4k})) = L(\pi\circ\beta) \leq L(\beta).$$

LEMMA 7.20. Let $X = (x,y) \in M^o$ and set $\tau(X) = d^M(P,X)$. Then:

(1) $|x| \leq \tau(X) \leq |x|(1 + O(|x|^{4k}))$;

(2) Whenever $X_1, X_2 \in U_X$, then

$$d^M(X_1,X_2) = |\pi X_1 - \pi X_2|(1 + O(|x|^{4k}));$$

(3) There exists $\tau > 0$ such that whenever $\tau(X) < \tau$, then

$$B(X,M;\tau(X)/2) \subseteq U_X.$$

Proof of (1). Let $\tilde{\gamma}$ be the path

(7.20.4) $\tilde{\gamma}(\sigma) = \sigma x$

from P to x in T. Then

$$\text{im}(\gamma \restriction (0,1]) \subseteq B_x,$$

where B_x is as in (7.18.5). There is a unique path γ from P to X such that $\pi\circ\gamma = \tilde{\gamma}$. (See Figure 14.) Now

(7.20.5) $\tau(X) \leq L(\gamma)$

$= L(\tilde{\gamma})(1 + O(|x|^{4k}))$, by Corollary 7.19.2

$= |x|(1 + O(|x|^{4k}))$.

Let γ_X be the geodesic from P to X, as in (3.4). Then

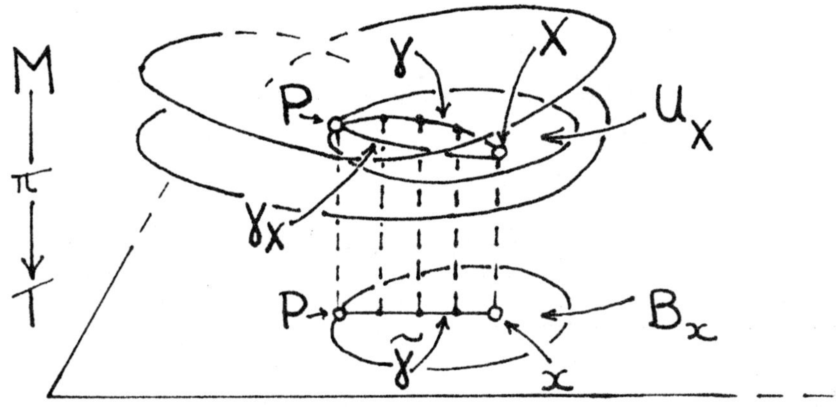

Figure 14

$$\begin{aligned} |x| &= |\pi \circ \gamma_X(1)| \\ &\leq L(\pi \circ \gamma_X) \\ &\leq L(\gamma_X), \qquad \text{by Corollary 7.19.2} \\ &\leq \tau(X). \end{aligned}$$

This together with (7.20.5) proves (1).

Proof of (2). Let $\tilde{\beta}$ be the path from $\pi(X_1)$ to $\pi(X_2)$ defined by

$$\tilde{\beta}(\sigma) = (1-\sigma)x_1 + \sigma x_2,$$

where, for $i = 1,2$, $X_i = (x_i, y_i)$. Then $\operatorname{im}(\tilde{\beta}) \subseteq B_x$. Let β be the path in U_X (see (7.18.6)) such that

$$\pi^{\circ} \circ \beta = \tilde{\beta}.$$

Then β must be a path from X_1 to X_2. Now

$$(7.20.6) \qquad \sup\{|\tilde{\beta}(\sigma)|\} \leq \sup\{|x'| \quad \text{for} \quad x' \in B_x\}$$
$$\leq 2|x|,$$

by (7.18.5). Hence

(7.20.7) $d^M(X_1,X_2) \leq L(\beta)$

$$= L(\tilde{\beta})(1 + O(|x|^{4k})),$$

by Corollary 7.19.2 and (7.20.6),

$$= |\pi X_1 - \pi X_2|(1 + O(|x|^{4k})).$$

On the other hand, consider a geodesic β^* from X_1 to X_2 in M; then

(7.20.8) $|\pi X_1 - \pi X_2| \leq L(\pi\circ\beta^*)$

$\leq L(\beta^*)$, by Corollary 7.19.2

$= d^M(X_1,X_2)$.

Now (2) follows from (7.20.7) and (7.20.8).

Proof of (3). Pick τ so small that in (1),

(7.20.9) $\tau(X) < 2|x|$ whenever $\tau(X) < \tau$.

Let $X_1 \in B(X,M;\tau(X)/2)$ and let β^* be a geodesic from X to X_1. Then for each σ,

$$|x - \pi\circ\beta(\sigma)| \leq L(\beta^*\restriction[0,\sigma]), \quad \text{by (7.20.8)}$$

$$\leq \tau(X)/2.$$

$$< |x|,$$

by (7.20.9). Thus $\text{im}(\pi\circ\beta^*) \subseteq B_x$. It follows that $\text{im}(\beta^*) \subseteq U_X$; in particular, that $X_1 \in U_X$. This proves (3). q.e.d.

(7.21) Proof of Theorem 7.18, concluded.

Proof of (1). This follows directly from the definitions.

Proof of (2). Let α and β be representatives of $\hat{\alpha}$ and $\hat{\beta}$ respectively. By Proposition 6.10,

(7.21.1) $|\hat{\alpha}| = \lim_{\sigma\to 0^+} d^M(P,\alpha\sigma)/\sigma,$

and

$$(7.21.2)\quad d^T(\hat{\alpha},\hat{\beta}) = \lim_{\sigma\to 0^+} d^M(\alpha\sigma,\beta\sigma)/\sigma < (1/4)|\alpha|$$

by hypothesis on $\hat{\beta}$. It follows from (7.21.1) and (7.21.2) that once σ is small enough,

$$d^M(\alpha\sigma,\beta\sigma) < (1/2)d^M(P,\alpha\sigma);$$

and so, by Lemma 7.20.3, $\beta(\sigma) \in U_{\alpha(\sigma)}$ for all sufficiently small σ. By Lemma 7.20.2, for all such σ,

$$(7.21.3)\quad d^M(\alpha\sigma,\beta\sigma) = |\pi\circ\alpha(\sigma) - \pi\circ\beta(\sigma)|(1 + O(|\pi\circ\alpha(\sigma)|^{4k})).$$

Now

$$|\pi\circ\alpha(\sigma)| = \sigma(|\alpha'0| + o(\sigma^0;\ \text{fixed } \alpha))$$

by (7.17.1); and

$$|\pi\circ\alpha(\sigma) - \pi\circ\beta(\sigma)| = \sigma(|\alpha'0 - \beta'0| + o(\sigma^0;\ \text{fixed } \alpha \text{ and } \beta))$$

for the same reason. Therefore (7.21.3) becomes

$$d^M(\alpha\sigma,\beta\sigma) = \sigma(|\alpha'0 - \beta'0| + o(\sigma^0;\ \text{fixed } \alpha \text{ and } \beta)).$$

Hence

$$\lim_{\sigma\to 0^+} d^M(\alpha\sigma,\beta\sigma)/\sigma = |\alpha'0 - \beta'0|.$$

Applying (7.21.2), this proves (2).

<u>Proof of (3)</u>. In view of (1) it is sufficient to prove (3) in deleted neighbourhoods of $\hat{P}$ in T_PM and of P in T. Let $x \in T^o$ be near P, and let $\tilde{\gamma}$ be as in (7.20.4). Since $\pi^o\colon M^o \to T^o$ is a p-fold covering map, there are exactly p distinct curves $\gamma_1,\dots,\gamma_p$ from P in M such that $\pi\circ\gamma_i = \tilde{\gamma}$, the

$\gamma_i \restriction (0,1]$ having images in the p different sheets of $(\pi^o)^{-1}B_x$. I claim that

(7.21.4) $\hat{\gamma}_1,\ldots,\hat{\gamma}_p$ are distinct in T_PM.

Suppose, for example, that γ_1 and γ_2 are equivalent. Then

$$d^M(\gamma_1\sigma,\gamma_2\sigma) = o(\sigma;\ \text{fixed } \gamma_1 \text{ and } \gamma_2);$$

while

$$\begin{aligned} d^M(P,\gamma_1\sigma) &\geq |\tilde{\gamma}\sigma|, \quad \text{by Lemma 7.20.1} \\ &= \sigma|x|, \end{aligned}$$

so that, once σ is small enough,

$$d^M(\gamma_1\sigma,\gamma_2\sigma) < (1/2)d^M(P,\gamma_1\sigma).$$

But then, by Lemma 7.20.3, for all sufficiently small σ, $\gamma_1(\sigma)$ and $\gamma_2(\sigma)$ are in the same sheet of $(\pi^o)^{-1}B_x$. This is a contradiction, which proves (7.21.4). Thus:

(7.21.5) $\hat{\pi}:(T_PM)^o \to T^o$ is everywhere at least p-to-one.

Now let $\hat{\alpha} \in (T_PM)^o$ be close to P, and set $x = \hat{\pi}(\hat{\alpha})$. Let $\tilde{\gamma}$ be as above, and let α be a representative of $\hat{\alpha}$. Then $\pi\circ\alpha$ and $\tilde{\gamma}$ are tangent at 0, by (7.17.1). Therefore

$$\begin{aligned} (7.21.6) \quad |\tilde{\gamma}\sigma - \pi\circ\alpha(\sigma)| &= o(\sigma;\ \text{fixed } \alpha) \\ &< (\sigma/2)|(\pi\circ\alpha)'0|, \end{aligned}$$

once σ is sufficiently small, because $|(\pi\circ\alpha)'0| = |\hat{\alpha}| \neq 0$ by (7.17.1) and Theorem 7.18.1. For such σ, say for $\sigma \in (0,\sigma_0]$,

$$|\tilde{\gamma}\sigma - \pi\circ\alpha(\sigma)| < (1/2)|\tilde{\gamma}(\sigma)|.$$

Now whenever $\sigma \leq \sigma_0$, in view of (7.20.4),

$$\begin{aligned} B(\tilde{\gamma}\sigma,T;|\tilde{\gamma}\sigma|) &\subseteq B(\tilde{\gamma}\sigma_0,T;|\gamma\sigma_0|) \\ &= B_{\tilde{\gamma}\sigma_0}, \end{aligned}$$

by (7.18.5). Hence

$$\pi \circ \alpha(\sigma) \in B_{\tilde{\gamma}\sigma_0}, \text{ for all } \sigma \in (0,\sigma_0].$$

Therefore $\operatorname{im}(\alpha \restriction (0,\sigma_0])$ is contained in one of the sheets of $(\pi^o)^{-1}B_{\tilde{\gamma}\sigma_0}$, which must be one of the $U_{\gamma_i\sigma_0}$, for $i = 1,\ldots,p$. By Lemma 7.20.2, for $\sigma \in (0,\sigma_0]$,

$$\begin{aligned} d^M(\alpha\sigma,\gamma_i\sigma) &= |\pi\circ\alpha(\sigma) - \tilde{\gamma}\sigma|(1 + O(|\tilde{\gamma}\sigma_0|^{4k})) \\ &= o(\sigma;\ \text{fixed } \alpha), \end{aligned}$$

by (7.21.6). That is, α is equivalent to γ_i; so $\hat{\alpha} = \hat{\gamma}_i$. Hence $\hat{\pi}:(T_PM)^o \to T^o$ is at most p-to-one everywhere. In conjunction with (7.21.5), this proves (3). q.e.d.

The problem remains of defining an exponential map at a singular point of a complex analytic hypersurface of dimension ≥ 2 under hypotheses more general than those of Theorem 7.6. In each of the following examples, M is the hypersurface in $\mathbb{C}^3$ defined as the zero-set of the given polynomial f, and P is the origin.

EXAMPLE 7.22. $f(x,y,z) = (x-y^2)^5 - z^2$. (See Figure 15.)

Here P is not an isolated singular point; the singular locus of M is

$$\Sigma = \{x = y^2,\ z = 0\}.$$

Condition (1.2) on an exponential map fails: let $Y = (y^2,y,0) \in \Sigma$ with $y \neq 0$; then there are at least two geodesics from P to Y in M.

Proof. Let α be a geodesic from P to Y, and write $\alpha(\sigma) = (x(\sigma),y(\sigma),z(\sigma))$. Then either

(i) $\alpha^*(\sigma) = (x(\sigma),y(\sigma),-z(\sigma))$ is a second geodesic from P to Y;

or

(ii) $\operatorname{im}(\alpha) \subseteq \Sigma$.

The path $\beta(\sigma) = ((\sigma y)^2,(\sigma y),0)$, when reparametrized by arc length, is a geodesic in Σ from P to Y, by symmetry of Σ. But β can definitely be shortened in M, for example by the path

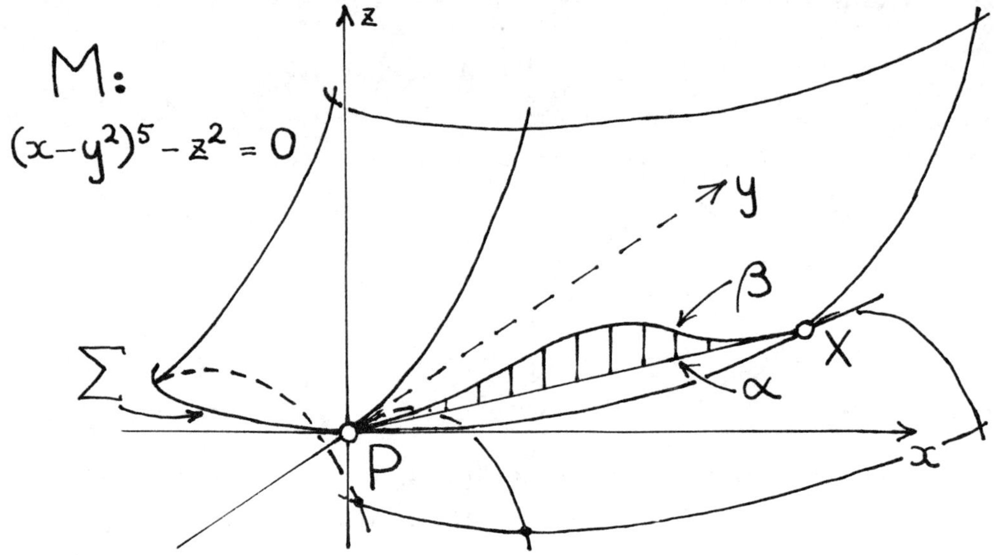

Figure 15

$$\beta^*(\sigma) = (\sigma y^2, \sigma y, (\sigma - \sigma^2)^{5/2} y).$$

Thus (ii) is false, so (i) musthhold. q.e.d.

EXAMPLE 7.23. $f(x,y,z) = x(y^3 - x^2) - z^2$. (See Figure 16.)

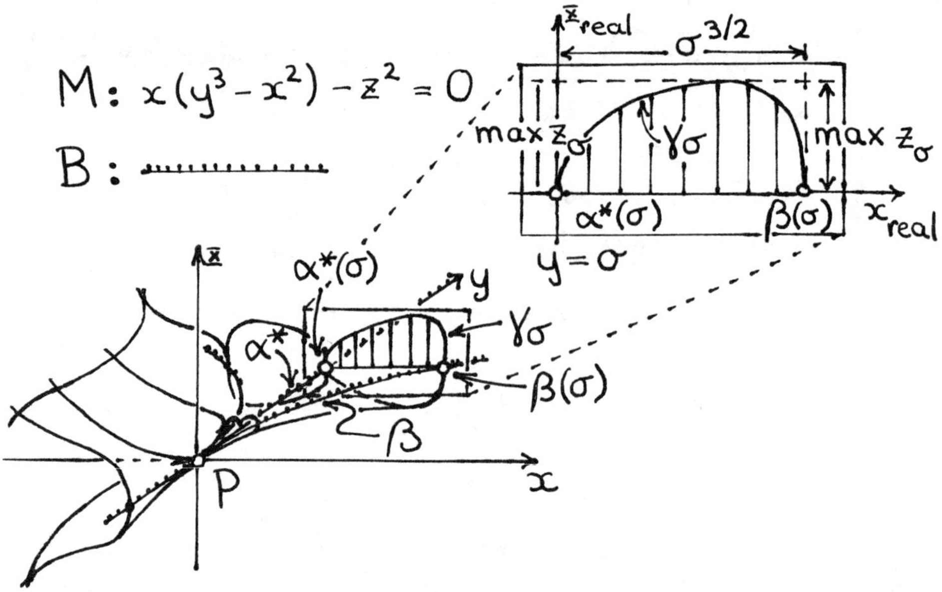

Figure 16

Here P is an isolated singular point, but not unbranched, and the branch locus B (see (7.1.8)) is not a cone; in fact

$$B = \{x^2 = y^3,\ z = 0\} \cup \{x = 0 = z\}.$$

In this example, even assuming the map e_P^{-1} of Chapter 3 exists, it is not one-to-one, so $\exp_P$ cannot be defined. Incidentally, f is a weighted homogeneous polynomial.

Proof. Let $\beta(\sigma)$ be the path $(\sigma^{3/2},\sigma,0)$, for $\sigma \in [0,1]$, and let α be the reparametrization of β by arc length. Then α is equivalent to β. Also α is a geodesic in B, by symmetry of B. If α is not a geodesic in M, then the argument used to establish Example 7.22 shows that (1.2) fails for M: any geodesic from P to $\alpha(1)$ can be reflected in the plane $\{z = 0\}$ to give a second such geodesic, since $M \cap \{z = 0\} = B$. Thus if α is not a geodesic then even e_P^{-1} fails to exist. So assume henceforth that

(7.23.1) α is a geodesic in M.

The path $\alpha^*(\sigma) = (0,\sigma,0)$ is also a geodesic in M. I claim that

(7.23.2) α^* and β are equivalent;

so that α^* and α are equivalent and e_P^{-1} (if it exists) cannot be one-to-one.

For each σ let γ_σ be the path in M from $\alpha^*(\sigma)$ to $\beta(\sigma)$ defined by

$$\gamma_\sigma(s) = (s\sigma^{3/2},\sigma,z_\sigma(s)), \qquad \text{where}$$
$$z_\sigma(s) = [s\sigma^{3/2}(\sigma^3 - s^2\sigma^3)]^{1/2}.$$

Regarding γ_σ as a curve in the real (x,z)-plane, it is concave downwards. For implicit differentiation yields

$$\frac{\partial z}{\partial x} = \frac{y^3-3x^2}{2x(y^3-x^2)} = \frac{1}{2x} - \frac{x}{y^3-x^2}\ ;$$

hence

$$\frac{\partial^2 z}{\partial x^2} = -\frac{1}{2x^2} - \frac{1}{y^3-x^2} - \frac{2x^2}{(y^3-x^2)^2}$$

which shows that $\gamma_\sigma''(s) < 0$ on $(0,1)$. It follows (see Figure 16) that

$$\begin{aligned} L(\gamma_\sigma) &\le \sigma^{3/2} + 2\max(z_\sigma) \\ &= \sigma^{3/2} + 2[(2/3^{3/2})\sigma^{9/2}]^{1/2} \\ &= O(\sigma^{3/2}). \end{aligned}$$

Therefore

$$d^M(\alpha^*(\sigma), \beta(\sigma)) = O(\sigma^{3/2}).$$

Thus α^* and β are equivalent, proving (7.23.2). q.e.d.

EXAMPLE 7.24. $f(w,x,y,z) = w^2 + x^2 + y^2 - h(z)$, where h is any analytic function (of z alone) of the form $h(z) = 3z^2$ + higher-order terms. Then M has an unbranched isolated singularity at P, but (1.6.3,VI) fails.

Proof. It is clear that P is an unbranched isolated singularity of M, so it remains to show that the sectional curvature function K^M is not bounded above on M^o. There exist $\delta > 0$ and a unique analytic function $z(t)$ of the form

$$z(t) = t(1 + O(t)),$$

defined for $|t| \le \delta$, such that for t real in $[0,\delta]$, the curve from P

$$\beta(t) = (t,t,t,z(t))$$

lies in M^o. Let $\underline{v}$ and $\underline{w}$ be the (constant) vector fields along β defined by

$$\underline{v}(t) = (1,-1,0,0), \quad \underline{w}(t) = (0,1,-1,0).$$

Then $\underline{v}$ and $\underline{w}$ are tangent to M^o, since they are both Hermitian-orthogonal to

$$\begin{aligned} \underline{\text{grad}}\ f &= (2\bar{w},\ 2\bar{x},\ 2\bar{y},\ \overline{h'(z)}) \\ &= (2t,\ 2t,\ 2t,\ \overline{h'z(t)}) \text{ at } \beta(t). \end{aligned}$$

Let $\Pi(t) \subseteq T_{\beta t}M$ be the 2-plane spanned by $\underline{v}(t)$ and $\underline{w}(t)$. I calculate $K^M(\Pi t)$ using (1.15.6) and (7.9.6). First note, however, that for any vector $u \in T_{\beta t}\mathbb{C}^4$,

$$\underline{\langle \underline{\text{grad}}\ f^{\perp}(u), \underline{wt} \rangle} = \underline{\langle u, \underline{wt} \rangle},$$

since $\underline{\langle \underline{\text{grad}}\ f, \underline{w} \rangle}(t) = 0$; in particular

$$\underline{\langle \underline{\text{grad}}\ f^{\perp}(D^{\mathbb{C}}_{\underline{v}}\ \underline{\text{grad}}\ f), \underline{w} \rangle} = \underline{\langle D^{\mathbb{C}}_{\underline{v}}\ \underline{\text{grad}}\ f, \underline{w} \rangle}.$$

Similaly for $D^{\mathbb{C}}_{\underline{w}}\ \underline{\text{grad}}\ f$. Moreover,

$$D^{\mathbb{C}}_{\underline{v}}\ \underline{\text{grad}}\ f = 2\underline{v},\quad D^{\mathbb{C}}_{\underline{w}}\ \underline{\text{grad}}\ f = 2\underline{w}.$$

Hence, from (7.9.6),

$$\begin{aligned}\underline{\langle R^M(\underline{v},\underline{w})\underline{v},\underline{w} \rangle} &= |\underline{\text{grad}}\ f|^{-2}\{\underline{\langle \underline{v}, D^{\mathbb{C}}_{\underline{v}}\ \underline{\text{grad}}\ f \rangle}\underline{\langle D^{\mathbb{C}}_{\underline{w}}\ \underline{\text{grad}}\ f, \underline{w} \rangle} \\ &\qquad - \underline{\langle \underline{v}, D^{\mathbb{C}}_{\underline{w}}\ \underline{\text{grad}}\ f \rangle}\underline{\langle D^{\mathbb{C}}_{\underline{v}}\ \underline{\text{grad}}\ f, \underline{v} \rangle}\} \\ &= 4|\underline{\text{grad}}\ f|^{-2}\{|\underline{v}|^2|\underline{w}|^2 - \langle \underline{v},\underline{w} \rangle^2\} \\ &= 4|\underline{\text{grad}}\ f|^{-2}|\underline{v} \vee \underline{w}|^2.\end{aligned}$$

Since this quantity is real,

$$\begin{aligned}K^M(\Pi t) &= \langle R^M(\underline{v},\underline{w})\underline{v},\underline{w} \rangle / |\underline{v} \vee \underline{w}|^2 \\ &= \underline{\langle R^M(\underline{v},\underline{w})\underline{v},\underline{w} \rangle} / |\underline{v} \vee \underline{w}|^2 \\ &= 4|\underline{\text{grad}}\ f(\beta t)|^{-2},\end{aligned}$$

which tends to $+\infty$ as $t \to 0$. q.e.d.

<u>Conjecture 7.24.1</u>. Let P be an unbranched isolated singular point of a complex analytic hypersurface M. Then M has an exponential map at P satisfying (1.1) -- (1.5).

I believe this conjecture for two reasons. First, to paraphrase Bochner [2], negative curvature is favored over positive even though the sectional curvature need not be bounded above. Bochner's results imply that the holomorphic sectional curvature and the Ricci curvature of M^o are everywhere ≤ 0. Second, M is a third-order perturbation of its Whitney tangent cone T at P. The method of proof of Theorem 7.18 was to regard M as being locally a perturbation of the x-axis; and similar methods might yield a proof of the present conjecture.

EXAMPLE 7.25. $f(x,y,z) = x(x^3 + y^3) - z^2$. P is an isolated singular point; it is not unbranched, but the branch locus B is a cone:

$$B = \{x(x^3 + y^3) = 0 = z\}.$$

There is no chart at P satisfying all the hypotheses of (1.6.3). It is still possible however (and, I believe, true) that M has an exponential map at P, even though Theorem 7.6 does not apply.

Proof. Fix $y \neq 0$ and set $Y = (0,y,0) \in B$. Near Y, x can be expressed as a holomorphic function of z,

$$x = (1/y^3)z^2 + \sum_3^\infty a_m z^m;$$

the coefficients a_m depend of course on y. The graph of this equation is a submanifold of M^o near Y of real dimension 2. Its Gaussian curvature K(y) does not depend on the a_m; in fact

(7.25.1) $K(y) = -8/|y|^3$.

Since B is a cone, $|y| = d^M(P,Y)$. If all the hypotheses of (1.6.3) held, Corollary 4.1.4 would give

$$K(y) = O(d^M(P,Y)^{-2}) = O(|y|^{-2});$$

but this contradicts (7.25.1). q.e.d.

Conjecture 7.26. Let M be a complex analytic hypersurface with an isolated singular point, P. Assume that the branch locus B at P (defined by (7.1.11)) is a cone from P. Then M has an exponential map at P satisfying (1.1) -- (1.5).

Assuming this conjecture holds, T_PM has a subcone T_PB isometric (under $\exp_P$) to B. Let T be the Whitney tangent cone of M at P; then B is a subcone of T. There is a natural map $\hat{\pi}$: $T_PM \to T$ defined as in (7.17); and $\hat{\pi}$ maps $T_PB \to B$ isometrically. The final conjecture generalizes Conjecture 7.10.7.

Conjecture 7.27. $\hat{\pi}$ restricts to a map

$$\hat{\pi}: \quad (T_PM - T_PB) \to (T - B)$$

which is a covering map and a local isometry.

PROBLEM 7.28. Let N be a Riemannian manifold. Let $M \subseteq N$ be a submanifold with an isolated singular point P; assume that M has an exponential map at P. Let f: $N \to N^*$ be a diffeomorphism of Riemannian manifolds. Does $M^* = f(M)$ have an exponential map at $P^* = f(P)$? If P is an isolated conical singularity of M is P^* likewise one for M^*? (Here M^o and $(M^*)^o$ have the Riemannian metrics induced from those on N and N^* respectively.)

A positive answer to the first question would imply the truth of Conjecture 7.24.1. It would also immediately extend the results of the present chapter to hypersurfaces in arbitrary complex manifolds.

BIBLIOGRAPHY

[1] A. D. Aleksandroff, Die innere Geometrie der konvexe Flaeche, Berlin, 1955.

[2] S. Bochner, Curvature in Hermitian metric, Bull. Am. Math. Soc., 53 (1947), 179-195.

[3] J. Cheeger, On the Hodge theory of Riemannian pseudomanifolds, Proc. Symp. Pure Math., v. 36, Amer. Math. Soc., Providence, R.I., 1980.

[4] J. Dieudonné, Foundations of Modern Analysis, Academic Press, N. Y., 1960.

[5] P. A. Griffiths, Complex differential and integral geometry and curvature integrals associated to singularities of complex analytic varities. Duke Math. J., 45 (1978), 427-512.

[6] D. Gromoll, W. Klingenberg and W. Meyer, Riemannische Geometrie im Grossen, Lecture Notes in Math., v. 55, Springer-Verlag, Berlin, 1968.

[7] P. Hartman, Ordinary Differential Equations, Wiley & Sons, N.Y., 1964.

[8] N. J. Hicks, Notes on Differential Geometry, Van Nostrand Reinhold, London, 1971.

[9] J. Milnor, Singular Points of Complex Hypersurfaces, Ann. of Math. studies, v. 61, Princeton Univ. Press, Princeton, N.J., 1968.

[10] D. Mumford, Algebraic Geometry I: Complex Projective Varieties, Grundlehren der math. Wissenschaften, v. 221, Springer-Verlag, Berlin, 1976.

[11] J. Nash, The imbedding problem for Riemannian manifolds, Ann. of Math. 63 (1956), 20-63.

[12] R. Palais, Morse theory on Hilbert manifolds, Topology 2 (1963), 299-340.

[13] D. Stone, Geodesics in piecewise linear manifolds, Trans. Amer. Math. Soc. 215 (1974), 1-44.

[14] D. Sullivan, Differential forms and the topology of manifolds, Proc. Tokyo Conf. on Manifolds, Univ. of Tokyo Press, Tokyo, 1973.

[15] B. Teissier, Cycles évanescents, sections planes et conditions de Whitney, Astérisque (Soc. Math. de France) 7 & 8 (1973), 285-362.

[16] R. Thom, Ensembles et morphismes stratifiés, Bull. Amer. Math. Soc., 75 (1969), 240-284.

[17] R. O. Wells, Jr., Differential Analysis on Complex Manifolds, Prentice-Hall, Englewood Cliffs, N.J., 1973.

[18] H. Whitney, Geometric Integration Theory, Princeton Univ. Press, Princeton, N.J., 1957.

[19] H. Whitney, Tangents to an analytic variety, Ann. of Math. 81 (1965), 496-549.

[20] H. Whitney, Complex Analytic Varieties, Addison-Wesley, Reading Mass., 1972.

General instructions to authors for

PREPARING REPRODUCTION COPY FOR MEMOIRS

For more detailed instructions send for AMS booklet, "A Guide for Authors of Memoirs." Write to Editorial Offices, American Mathematical Society, P. O. Box 6248, Providence, R. I. 02940.

MEMOIRS are printed by photo-offset from camera copy fully prepared by the author. This means that, except for a reduction in size of 20 to 30%, the finished book will look exactly like the copy submitted. Thus the author will want to use a good quality typewriter with a new, medium-inked black ribbon, and submit clean copy on the appropriate model paper.

Model Paper, provided at no cost by the AMS, is paper marked with blue lines that confine the copy to the appropriate size. Author should specify, when ordering, whether typewriter to be used has PICA-size (10 characters to the inch) or ELITE-size type (12 characters to the inch).

Line Spacing – For best appearance, and economy, a typewriter equipped with a half-space ratchet – 12 notches to the inch – should be used. (This may be purchased and attached at small cost.) Three notches make the desired spacing, which is equivalent to 1-1/2 ordinary single spaces. Where copy has a great many subscripts and superscripts, however, double spacing should be used.

Special Characters may be filled in carefully freehand, using dense black ink, or INSTANT ("rub-on") LETTERING may be used. AMS has a sheet of several hundred most-used symbols and letters which may be purchased for $5.

Diagrams may be drawn in black ink either directly on the model sheet, or on a separate sheet and pasted with rubber cement into spaces left for them in the text. Ballpoint pen is *not* acceptable.

Page Headings (Running Heads) should be centered, in CAPITAL LETTERS (preferably), at the top of the page – just above the blue line and touching it.

LEFT-hand, EVEN-numbered pages should be headed with the AUTHOR'S NAME;

RIGHT-hand, ODD-numbered pages should be headed with the TITLE of the paper (in shortened form if necessary).

Exceptions: PAGE 1 and any other page that carries a display title require NO RUNNING HEADS.

Page Numbers should be at the top of the page, on the same line with the running heads.

LEFT-hand, EVEN numbers – flush with left margin;

RIGHT-hand, ODD numbers – flush with right margin.

Exceptions: PAGE 1 and any other page that carries a display title should have page number, centered below the text, on blue line provided.

FRONT MATTER PAGES should be numbered with Roman numerals (lower case), positioned below text in same manner as described above.

MEMOIRS FORMAT

It is suggested that the material be arranged in pages as indicated below.
Note: Starred items (*) are requirements of publication.

Front Matter (first pages in book, preceding main body of text).

Page i – *Title, *Author's name.

Page iii – Table of contents.

Page iv – *Abstract (at least 1 sentence and at most 300 words).

*1980 Mathematics Subject Classifications represent the primary and secondary subjects of the paper. For the classification scheme, see Annual Subject Indexes of MATHEMATICAL REVIEWS beginning in December 1978.

Key words and phrases, if desired. (A list which covers the content of the paper adequately enough to be useful for an information retrieval system.)

Page v, etc. – Preface, introduction, or any other matter not belonging in body of text.

Page 1 – Chapter Title (dropped 1 inch from top line, and centered).

Beginning of Text.

Footnotes: *Received by the editor date.
Support information – grants, credits, etc.

Last Page (at bottom) – Author's affiliation.

ABCDEFGHIJ – AMS – 898765432